BEYOND

Our Future in Space

Chris Impey

W. W. NORTON & COMPANY

New York   London

For information about permission to reproduce selections from this
book, write to Permissions, W. W. Norton & Company, Inc.,
500 Fifth Avenue, New York, NY 10110

For information about special discounts for bulk
purchases, please contact W. W. Norton Special Sales
at specialsales@wwnorton.com or 800-233-4830

Manufacturing by RR Donnelley, Harrisonburg
Book design by Ellen Cipriano
Production managers: Devon Zahn and Ruth Toda

Library of Congress Cataloging-in-Publication Data

Impey, Chris.
Beyond : our future in space / Chris Impey. — First edition.
pages cm
Includes bibliographical references and index.
ISBN 978-0-393-23930-0 (hardcover)
1. Astronautics—Forecasting—Popular works. 2. Astronautics—
History—Popular works. 3. Manned space flight—Popular works.
4. Outer space—Exploration—Popular works. I. Title.
TL793.I468 2015
629.401'12—dc23

2014046924

W. W. Norton & Company, Inc.,
500 Fifth Avenue, New York, N.Y. 10110
www.wwnorton.com
W. W. Norton & Company Ltd., Castle House,
75/76 Wells Street, London W1T 3QT

1 2 3 4 5 6 7 8 9 0

To the next generation of space pioneers,
who will realize their dreams to
live off-Earth.

CONTENTS

Preface

Space is inhuman. We can't survive unprotected in a pure vacuum for more than a minute. Going there involves being strapped to a barely controlled chemical explosion. Low Earth orbit is equivalent to half an hour's drive straight up, but it's insanely expensive to get there. The set of people who have experienced zero gravity is one of the most exclusive clubs in history. Yet space travel is an expression of a fundamental human trait: the desire to explore.

This book is an exploration of the past, present, and future of space travel. We're on the cusp of an important transition, where diverse technologies have matured to the point where space travel could be routine. A cadre of innovators and entrepreneurs is about to deliver space travel that's not just for astronauts and the super-rich. It will happen sooner than you think.

Each of the four parts of this book is preceded by a fictional vignette delivering us into the world of a young pioneer about to undertake a journey to the stars. First, we look at the "Past" to learn about our genetic proclivity for exploration and our progress with rocketry that let us leave the Earth for the first time in the mid-twentieth century. We also learn about the highs and the lows resulting from that landmark achievement. Then, in the "Present," we see that the malaise of the space program will be cured by a new generation of entrepreneurs who are transforming our potential to leave the planet. We examine the legal and regulatory barriers that stand in their way and consider the dangers faced by space travelers. Peering into the "Future," we look at how we can travel to the Moon and Mars, and we investigate the technologies needed to establish colonies there. We meet the robots that will be our partners in space and we visualize a time when off-Earth humans become a new species. Finally, we speculate "Beyond" our current capabilities to the time when we can travel to the stars and become citizens of the Milky Way. In a universe built for life, our yearning for cosmic companionship is strong, and we may never realize our full potential as a species if we stay Earthbound.

I'm grateful to Anna Ghosh for stalwart support of my writing career, and Tom Mayer and Ryan Harrington at W. W. Norton for helping to forge a better book. I've benefited from conversations with many colleagues over the years, but any errors or misstatements that remain are my responsibility alone. My deepest gratitude goes to Dinah Jasensky for her love and encouragement of all my writing endeavors.

PART I

PRELUDE

I was four years old when I was chosen to be a Pilgrim. Too young to understand what that meant, I heard my mother try to describe it, but her words made no sense, so I fixated instead on the mixture of excitement and fear in her voice.

A few years later, the implications came into better focus. I suffered difficult years in a normal school, where I was alternately ostracized and bullied, ignored and humiliated, with a relentlessness seen only in children. To be a Pilgrim was an extraordinary honor but it marked me as different, as other. It was a relief when I was plucked from there at age eight and sent to a special school called the Academy, a school for my kind.

The Academy was in Switzerland, on a mountain lake that sparkled aquamarine in the sunlight, in a special compound where we were shielded from the media and prying eyes. There were nearly 300 of us, from more than fifty countries. We got one trip home every year but no family member could visit the school. Video chats with outsiders were limited to an hour each week. It may sound harsh, but it was for our own good.

We ranged in age from seven to twelve. I was one of the youngest. There were equal numbers of boys and girls. With so many cultures and languages represented, it could have been like Babel, so most of us wore our digital translators all the time. The curriculum was also polyglot. We studied everything from engineering and philosophy to medicine and fine arts. With as many tutors and counselors at the Academy as students, we had plenty of help. Expectations were high and there was a gentle but persistent pressure to excel. The staff was aloof, no doubt instructed not to form emotional bonds with us. Some of them were psychologists and psychiatrists who were clinical and often cold in their detachment.

I remember my dreams from that time.

Large shapes moving around in the darkness. An unrelenting pressure in my chest. A door sliding up, an inch from my face, like the lid of a coffin. A window, and beyond it, nothing, an absolute void. The images were both inchoate and sharply real. I would always wake with a start and sit up, bathed in sweat, my breathing fast and shallow.

On my mother's last visit before I left the Academy, a few months after I turned seventeen, she told me how my father died. I never knew the details when I was young, and information at the Academy was tightly controlled. He was on his second tour of a mining station on Phobos. His crew was deep in a shaft looking for inclusions of platinum and iridium. Miscommunication with the surface crew caused them to set off a charge nearby. Sofa-sized slabs of rock were ejected from the face of the shaft. At the center of a small asteroid there's no gravity, so no way to be crushed by falling rock as on Earth.

But Newton's laws have their own implacable logic. A large rock hit him squarely in the chest and carried him to the opposite wall, where its momentum was transmitted into his body, crushing and killing him instantly.

That's probably part of why I was chosen. My father was a space rat and my sister an accomplished pilot; I had space in my genes. But at the Academy there were many kids with no predisposition to technical subjects; their parents were musicians or artists or diplomats. No two of us were alike. We seemed to form, as intended, a miniature world.

At my graduation, my mother and my sister were in the auditorium, and they grinned ear to ear when I waved to them. Later, as we shared a meal in the dining hall over-looking the lake, I flashed back to the way my mother's voice had sounded a decade earlier. Her excitement and fear had morphed into quiet pride tinged with sadness.

After graduation, we had two weeks to pack up and say our farewells. Visits and video chats were unlimited. I remember that it was emotionally exhausting to spend so much time with my family. Many of the other students felt the same way. We were young adults and had grown up without them, with only each other for sustenance. I felt relieved when the time came to travel to the launch site, though that relief was followed swiftly by a wave of guilt.

The time had come to learn what it meant to be a Pilgrim.

To be an emissary of Earth in the late twenty-first century. To be in a small group chosen for a unique experiment. The experiment was designed by sober scientists and engineers but it had the trappings of madness. We were human seedlings, charged with taking root in a new world.

1

Dreaming of Beyond

Out of Africa

When we were just one million strong, did we dream about what lay beyond?

Two hundred thousand years ago, anatomically modern humans first emerged in Northeast Africa.[1] The cradle of our creation was the place now known as Ethiopia. Over the next hundred thousand years, these humans spread across Africa. Our distant ancestors kept no journals and, as far as we know, they had no written language. Only bones and scattered artifacts survive. Those artifacts speak of a rugged, doughty species that ceremonially buried their dead, hunted with sharpened flints made into spears and arrows, and daubed paint on cave walls to record the iconography of their lives. Their evocative images, which must have seemed kinetic in the flickering glow of an oil lamp or a fire, speak to us across the millennia of their fears and dreams.

Modern genetic techniques have allowed us to reconstruct their journey out of Africa—an epic migration as audacious as our first steps into space many millennia later.

Life on Earth is united by a single genetic code. A four-letter alphabet of base pairs encodes the unique function and form of every organism. The four bases—A for adenine, C for cytosine, G for guanine, and T for thymine—form the rungs of the twisted ladder that is DNA. A pairs with T and C pairs with G across the ladder; when the ladder splits down the middle, each side is the template for a new DNA molecule. The genetic code specifies the sequence of twenty amino acids used by living cells to build proteins.

If the genetic code were perfectly transcribed and expressed, there would be no evolution and life would be, well, boring. It would also be a dead end, since it couldn't adapt and survive over time. One type of variation occurs when the genetic blueprint, the genotype, is expressed in a particular environment, the phenotype. Two cloned seedlings will develop quite differently when one grows in loamy soil and the other grows on a windswept mountain. A second type of variation occurs over time when the genetic material is altered by mutation or imperfect copying. Biodiversity cascades as variations grow over time and are culled by natural selection. As a result, the DNA of life has developed a tangle of branches that emerge from the root, a "last common ancestor" four billion years ago.[2] This primitive cell was the precursor of all plants and animals and the mother of all microbes. Creatures that reproduce sexually, like humans, mix their DNA in a way that makes everyone unique from their parents. This accelerates genetic variation and evolution, particularly in small populations.[3]

Genetic anthropology is the use of DNA and physical evidence to trace human migration. We all contain DNA from our last common ancestor, which is how we know when and where humans originated. DNA mixes due to sexual reproduction but some special sequences of DNA pass unaltered from parent to child. For example, the Y chromosome passes only from father to son and so allows men to trace paternal lineages, while mitochondrial DNA passes only from mother to child and so allows both men and women to trace maternal lineages. Both of these sequences of DNA are subject to occasional harmless mutations

that become inheritable genetic "markers." Within
region, any particular genetic marker spreads quic
generations it's found in almost every member of
When people migrate from a region, they carry th
By studying different genetic markers in many ind
scientists map out early human migration.

The Genographic Project has painted a picture of human migration
using "brushstrokes" of DNA from more than 70,000 members of carefully selected indigenous tribes around the world. Appropriately, most
of the funding comes from the National Geographic Society, which has
turned from exploration of the planet to exploration of the inner world.
The project is not without controversy, as some indigenous peoples have
declared it exploitative and have declined to take part. However, the
project has gained a big boost from crowdsourcing. More than 600,000
people have received their genetic histories in return for contributing
DNA to an open-source database.[4] With such a rich resource and powerful computers to apply to it, more than 100,000 genetic markers have
been identified in the past decade. The leader of the project is Spencer Wells, a National Geographic explorer-in-residence. He says: "The
greatest history book ever written is the one hidden in our DNA."

Our DNA tells the story of the profound human urge to explore.

Around 65,000 years ago, we first ventured out of the continent of
our origin. The route from the Horn of Africa to the Arabian Peninsula
was probably across the Bab el-Mandeb strait. Today that strait is one
of the world's busiest shipping lanes; at that time, after the last ice age
had lowered sea levels, it was merely a narrow, shallow channel. The
tribe that ventured out of Africa may have been only a few thousand
strong. It was not a single expedition but a series of small clans of loosely
related family members leaving over a period of centuries. They prospered as they dispersed, starting settlements in Central Asia and then
in Europe. By 50,000 years ago, they had spread to southern China and
Australia. By 40,000 years ago, they'd spread throughout Europe. Pop-

ons prospered thanks to hospitable conditions in southern Europe and Asia.

The last stage of the migration was audacious and dramatic. Despite more favorable climates around the Mediterranean and in the Middle East, some nomads ventured northward. The most recent ice age was sharpening, but these intrepid humans spread in an arc across the Siberian tundra. Vast ice sheets had sucked much of the moisture out of the Earth's atmosphere and dropped sea levels hundreds of feet. This allowed our ancestors to traverse the land bridge across the Bering Strait about 16,000 years ago. There's evidence they reached southern California just 3,000 years later. It took them only another few thousand years to travel south most of the way through the Americas. Looking at a map of the journey, where our ancestors moved from the frozen wastes of Alaska to the bleak landscape of Patagonia, it seems astonishing that they traveled so swiftly—the migration couldn't have been motivated simply by food and shelter.

The timeline just described is affected by the possibility that humans migrated by sea. There is some indication that small groups made the arduous voyage across the Atlantic from Europe to North America 25,000 years ago, clinging to the edge of the ice pack. In Australia, a single lock of Aboriginal hair is rewriting the story of how that continent was populated. The traditional explanation is that some humans who had left Africa moved east and settled in Australia after a sea voyage from Southeast Asia. But in 2011, gene sequencing of hair donated to a British anthropologist in 1923 showed that Aboriginal Australians are more closely related to Africans than they are to Europeans or Asians. So present-day native Australians may be the oldest group of humans living outside Africa.[5]

After tens of thousands of generations on the African savanna, we spread across the Americas in a few hundred (Figure 1). This rapid, purposeful exploration of new worlds is in its way as dramatic in terms

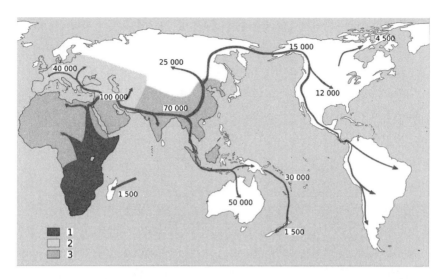

Figure 1. Map of early human migrations, based on DNA in mitochondrial genomes. The migration routes are marked in years before the present day. Different shadings are for Homo sapiens (1, dark gray), early hominids (3, mid gray), and Neanderthals (2, light gray).

of leaving our comfort zone and embracing the unknown as our decision to leave the Earth when we developed the technology to do so.

Genetic material can tell us *how* we spread around the world, but it can't tell us *why*. For that, we have to look into our natures.

The Urge to Explore

Epic animal migrations are driven by climate, the availability of food, or mating, and almost all animal migrations are seasonal. Humans are the only species that moves systematically and purposefully over very large distances, in multigenerational migrations, for reasons not tied to the availability of resources. The itch that led our ancestors to risk everything to travel in small boats across large bodies of water like the Pacific Ocean is related to the drive that will one day lead us to colonize Mars. Its origins lie in a mixture of culture and genetics.

Behavioral psychologist Alison Gopnik has observed that humans are unique in the way they connect play and imagination. Mammal species can be playful when they're young, but the play is quickly channeled into practicing skills such as hunting and fighting, which are needed as an adult. Human children spend a proportionally longer time in a world where their development is sheltered and facilitated by adults.[6] We play, according to Gopnik, by creating hypothetical scenarios that allow us to test hypotheses—acting in effect like miniature scientists. What happens if I mix these two liquids together? If I go through the woods, will I be able to remember enough landmarks to find my way back? Can I make my Lego bridge span the gap from the sofa to the coffee table? Children are fearless hypothesis machines. After the child develops the necessary motor skills, mental exploration then leads to investigation of the physical environment.

The development of hypothetical scenarios through play isn't needed for survival and the tendency for mental exploration is peculiarly human. Restlessness isn't only in our minds; it's also in our genes.

We share more than 95 percent of our DNA with monkeys and apes, so we have great commonality with our most recent ancestors. Yet certain developmental genes gave us an edge over apes and other hominids: We have lower bodies built for walking long distances, hands that are better for manipulating objects, and brains with larger language and cognition regions. These genes are regulated by regions of DNA that used to be labeled "junk" but are now recognized as being keys to understanding how a species evolves.[7]

One particular gene has received a lot of attention because of its central role in controlling one of the most important neurotransmitters. DRD4 is one of the genes that control dopamine, a chemical messenger that influences motivation and behavior. People with one of the variants of this gene, called 7R, are more likely to take risks, explore new places, seek and crave novelty, be extroverts, and be hyperactive. About one person in five carries DRD4 in the 7R form.

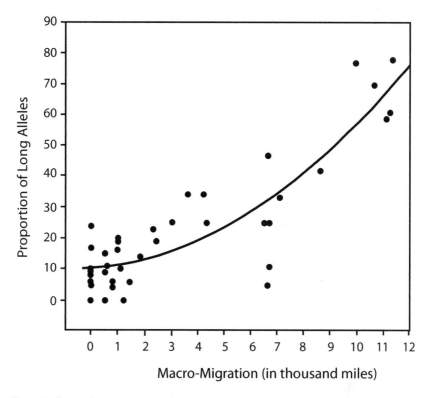

Figure 2. Correlation between the frequency of DRD4 alleles and long-distance migration among 39 population groups over the past 30,000 years. In modern populations, the long or 7R variant of this gene is associated with attention deficit hyperactivity disorder (ADHD).

Intriguingly, the 7R mutation probably first occurred about 40,000 years ago, soon after the exodus from Africa, when humans began fanning out across Asia and Europe. Other studies explicitly tie 7R to migration. Research by Chuansheng Chen at the University of California Irvine showed that among the largely stationary populations of Asia, only 1 percent currently have 7R, while the prevalence is 60 percent in present-day South Americans, whose populations traveled enormous distances from Asia beginning 16,000 years ago (Figure 2).[8]

So is there an "exploration gene"? No. Genes work in combination with each other and behavior is sculpted by the environment, so genes are not destiny and no single gene can hardwire us for exploration. Also, unknown situations can be fraught with danger, so a gene that spurs exploration doesn't necessarily offer a selective advantage. Moreover, when this gene is expressed, it can have a downside. People with the 7R variation are two and a half times more likely to suffer from ADHD, 50 percent more likely to be sexually promiscuous (which is culturally frowned upon yet is actually an evolutionary advantage . . .), and they're prone to alcoholism and drug addiction. The safe functioning of any hunter-gatherer society requires intensive cooperation and stable social relationships; too much thrill-seeking would be dangerous and disruptive.

However, in situations of resource scarcity or stress, this particular mutation shows its advantages. Carriers of 7R not only are comfortable with change, they also startle less easily.[9] They use less emotion in making decisions and they're less impacted by the negative emotions of others. Low emotional reactivity and high emotional endurance are valuable traits for a human in a perilous, new environment, as is the ability to plan and solve complex problems when faced with a threat. This "adventure" genotype may even protect against stress, anxiety, and depression.

Even if they're only present in a fraction of the population, the traits that favor adventurousness are self-reinforcing. If the 7R mutation has slightly higher frequency in a population that migrates, that frequency will increase in a finite gene pool. Mobility and dexterity are enhanced as they are expressed. The most successful nomads will encounter new sources of food and new possibilities for enhancing their lifestyle. The best users and makers of tools will be spurred to come up with new tools and novel applications of existing tools. The fulcrum of this feedback loop is our one attribute that's unparalleled: a big brain.

Making Mental Models

What would it be like to be a dog?

Despite the empathetic bond that connects us with our "best friends," a species gulf prevents us from answering that question. Dog brains resemble ours in structure and they undergo chemical changes associated with a wide range of emotional states. Like us, dogs can dream. There's also intriguing evidence that dogs can mentally sort objects into categories, a talent for abstract thought only previously demonstrated among certain primates and birds. But a dog's emotional development stops at a level corresponding to the maturity of a human toddler.

Dogs are unable to make mental models of the sort that humans thrive on. If you were suddenly trapped inside the mind of your dog, you'd be subjected to a cacophony of smells and visual stimuli, adept at molding your behavior according to the external environment and your owner's wishes. But you'd never make mental models based on your experience to guide you in future decisions.

Humans have a singular ability to reason with language and symbols. It starts early. Between the ages of six and nine months, a baby will move from babbling and mimicry to attaching words to real objects. Around the same time, a baby becomes able to hold the idea of an object even when the object is removed from view. Both transitions involve the creation of a mental scheme as a proxy for the real world.

By the age of two, a child is able to detect statistical patterns and draw inferences about cause and effect from that evidence. In an example reported by Alison Gopnik, two-year-olds were faced with a toy box containing green frogs plus a few yellow ducks.[10] The experimenter took a few toys from the box, seemingly at random, and asked the baby for one. The baby showed no color preference if the experimenter drew green frogs from the box of mostly green toys. But the baby specifically gave the experimenter a duck if the experimenter had drawn the relatively rare yellow ducks from the box. The baby knew it was unlikely for the experimenter to draw mostly ducks, so the experimenter's behavior indicated a preference for ducks. Babies aren't doing experiments or crunching statistics in the self-conscious way that adults do, but they're unconsciously processing information in a way that parallels the scientific method.

The next level of development involves play. When children say, "Let's pretend," they conjure up alternative worlds and populate them with imaginary friends. As we all know, these imaginary worlds can be very elaborate. Such behavior is uniquely human. Jane Goodall only spotted a few examples of pretend play in many hours of observing the Gombe chimpanzees in Tanzania, while it would be trivial to note this behavior in any four-year-old. Conceptually, children engage in counterfactual thinking—speculation beyond normal experience. This is a step beyond making mental models of the actual world to making mental models of strange and unfamiliar worlds. Scientists use counterfactual thinking as a high-level skill for developing theories, by asking, "What might be but isn't, and why?" Children use it to develop skills that will let them explore the environment they will live in as adults.

We've seen that there's a tight connection between play and abstract, logical thinking in children. Psychologists used to believe that all reasoning involved the use of logic. But real life is usually so messy that rules such as "if this, then that" don't apply. Logic requires premises and assertions that may be difficult to test. Worse, there may be a huge number of logical inferences from a particular set of assertions, and valid conclusions can be drawn even when they conflict with facts. Human reason isn't neat or simple, and it bears little resemblance to a proof in logic.

A better description of reasoning is the making of mental models to represent phenomena we encounter. A mental model is like a simulation of some aspect of the world fleshed out by our knowledge and informed by our experience. It's a dynamic process in which we readily adapt or discard models that are found wanting.[11] We test our models with hypothetical situations, making a multitude of mental models for all the possible situations we might face. It's easy and cheap—we don't need any equipment or tools and we don't need to put ourselves in danger. "If I climb out on the tree branch to get the honey, the branch might break and the bees might sting me. But if I snag the branch with a vine, I can break it and come and get the honey when the bees have dispersed." Everything happens in our heads!

Scholars argue vigorously about when abstract thought and reasoning first emerged. After all, it's hard enough to know what your spouse or best friend is thinking, let alone an early ancestor who's been dead for a hundred thousand years.

There's wide agreement that humans were anatomically modern about 200,000 years ago. Until a few decades ago, the conventional wisdom dated creative, abstract thought to about 40,000 years ago, after humans left Africa and radiated across Europe and Asia and around the same time developed the 7R mutation. Scientists date the earliest examples of cave paintings, as well as bones and stones carved into artwork and tools, to about this period. The relatively sudden emergence of language and modern human behavior has been called the "Great Leap Forward." Attributes that make us modern are summarized by the American neurophysiologist William Calvin as the behavioral *b's*: blades, beads, burials, bone tools, and beauty.[12] The last item implies aesthetic judgment and forms of representation that include playing games, telling stories, and creating art and music.

But recent discoveries have cast doubt on this hypothesis. In a cave above the windswept coast of South Africa, archaeologists found an abalone shell containing a dried paste made from charcoal, iron-rich dirt, crushed animal bones, and an unknown liquid. The shell was a prehistoric paint can. In eastern Morocco, carved and painted shells were found that had been used as decorative beads. Elsewhere in Africa are several sites where complicated animal snares and traps were dis-covered. All three types of artifact date from about 80,000 years ago, and there are even earlier hints of abstract thinking. This evidence points to a gradual accumulation of knowledge, skills, and culture over several hundred thousand years, rather than a "Great Leap Forward."

Regardless of *when* we evolved these uniquely human capabilities, renowned psychologist Steven Pinker put his finger on a problem, the problem of *why*. He wonders, "Why do humans have the ability to pur-sue abstract intellectual feats such as science, mathematics, philosophy, and law, given that the opportunities to exercise these talents did not exist in the foraging lifestyle in which humans evolved, and would not have parlayed themselves into advantages in survival and reproduction even if they did?"[13] In other words, in our modern culture math and sci-ence get applied in myriad ways to help us master our environment and live longer, but abstract concepts have no utility to a hunter-gatherer who must find food and shelter daily.

Pinker and other evolutionary psychologists speculate that these traits emerged as a by-product of natural selection. In this view, we occupy a "cognitive niche" in evolution not shared by any other spe-cies.[14] We manipulate the environment and form complex social net-works so we can meet environmental challenges quickly, while animals use the much slower process of genetic evolution. For example, our dexterity let us create new tools and hunting strategies that were most effective when used cooperatively. Meanwhile, our language permitted the evolution of sophisticated behaviors like altruism and reciprocity, and mental models let us efficiently play out hypothetical scenarios before seeing how they operate in the real world. An adaptation useful for one purpose might prove to be useful for another. For example, meat is a concentrated source of nutrients for an opportunistic omnivore, but bringing down an animal takes more smarts than gathering berries, so eating meat would facilitate greater intelligence. Our social, mental, and physical capabilities coevolved. It's a difficult idea to prove, but it does explain how abstract reasoning could evolve even if it played no immediate role in survival.

Many Worlds

Imagine being transported back in time 2,500 years and across the world to a bustling port on the Ionian coast, into the company of a philosopher called Anaxagoras (Figure 3). An intense and austere young man, Anaxagoras thought that the opportunity to understand the universe was the reason why it was better to be born than not to exist.

Anaxagoras was part of a wave of thinking that held that the Earth was not unique among the many worlds in space.[15]

To see why this was bold, think about how we regarded the sky for many millennia up until the time of the ancient Greeks. The sky was a map, a clock, a calendar, and a repository of myth and legend. It seemed obvious that the Earth was stationary and at the center of the universe

Figure 3. Greek philosopher Anaxagoras, from a fifteenth-century manuscript called *Nuremberg Chronicles*. Anaxagoras lived from 510 to 428 BC and was the first philosopher to propose a natural mechanism for the cosmos and embrace the idea of pluralism or "many worlds."

while the heavens wheeled overhead. The Sun, the Moon, the planets, and the stars were remote and inaccessible. Prehistoric humans had the ability to think abstractly and make mental models, but there's no sign they used this ability to conceive of what lies beyond the Earth.

Anaxagoras moved from Ionia to Athens, where he gravitated toward the center of intellectual life. The great Greek playwright Euripides incorporated Anaxagoras's theory of mind into his tragedies, and his friend Pericles became the greatest statesman and orator of the Golden Age of Athens. Anaxagoras was prolific in his novel ideas and revolutionary theories. He believed that the Sun was a mass of molten metal much bigger than the Peloponnese peninsula, the Moon was a rock like the Earth that didn't emit its own light, and the stars were fiery stones. He thought that the Milky Way represented the light of countless stars. He gave physical explanations for the Sun's seasonal motion, the motions of the stars, eclipses, and the origin of comets. Anaxagoras speculated that the cosmos was originally undifferentiated but contained all its eventual constituents. He saw no logical limits to the formation of structure, so he proposed that there can be endless worlds within worlds, either large or small.[16]

Original ideas tend to provoke major backlash. Anaxagoras was charged with atheism for proposing a physical and natural explanation for the universe, with no reference to gods or divine intervention, and for daring to suggest that the Sun was as large as Greece. His close friendship with Pericles hurt him because the famous politician had powerful enemies. He avoided death by escaping back to Ionia, where he spent the rest of his life in exile.

The idea of "many worlds" had an antecedent in the work of Thales, often called the father of philosophy, who conjectured that space was infinite, and it continued in the work of the Atomists and Epicureans. Just a few centuries later, the Roman philosopher and poet Lucretius boldly wrote, "In the universe, nothing is only of its kind. In other regions, surely there must be other Earths, other men, other beasts of burden."[17]

Greek philosophy sought to replace fear and superstition with rational thought. Humans had long had the capacity for abstraction, but in the hands of the Greeks it was augmented with mathematics and formal rules of logic. Aristarchus used geometry and an understanding of eclipses and lunar phases to deduce that the Sun must be larger than the Earth, and this led him to propose a heliocentric model nearly two thousand years before Copernicus. Eratosthenes combined his knowledge that the Earth was round—from the shape of its shadow in a lunar eclipse—with the way the Sun casts shadows at different places on the Earth's surface, to estimate the size of the Earth. This philosopher, who had never traveled more than a hundred miles in his life, could understand what was unknown to the early humans who had made epic migrations across the planet.

The philosophers of ancient Greece extended mental models into entirely new regimes. Democritus asked what would happen if he took a sharp knife to a stone and subdivided it over and over again. Logically, the process can continue forever to the infinitely small, or it can end with a fundamental, indivisible unit of matter. He rejected infinitely small particles and hypothesized atoms. Archytas asked what would happen if he went to the edge of space and hurled a spear outward. If the spear met a barrier, it would beg the question of what lay beyond the barrier, so he suggested that space is infinite. Without atom-smashers and telescopes, the Greeks couldn't resolve these questions, but their "thought experiments" heralded the birth of science.[18]

Unfortunately for the progress of cosmology, the many-worlds idea was squelched by the countervailing view and intellectual dominance of Aristotle, who believed the Earth was unique and there could be no system of worlds. Aristotle's Earth-centric view took root because it agreed with common sense—if the Earth was not the center of the universe, it would be moving rapidly, and no motion was apparent. The geocentric cosmology was built into Christian theology, where it was consistent with a special relationship between humans and their Creator. Meanwhile, other religious traditions were more sympathetic to the idea of many worlds. Hinduism and Buddhism preach a multiplicity of worlds with intelligent life inhabiting them.[19] In one myth, the god Indra says, "I have spoken only of those worlds within this universe. But consider the myriad of universes that exist side by side, each with its own Indra and Brahma, and each with its evolving and dissolving worlds."

Powerful ideas grip the imagination. Although the Western intellectual tradition was inhospitable, poets and dreamers speculated about what lay far beyond the Earth. In the second century AD, Lucian of Samosata wrote a romantic fantasy that anticipates modern science fiction.[20] "A True Story" told of people transported to the Moon, where they encounter a humanlike race riding on the backs of three-legged birds. In this robust fantasy, people and fantastical creatures populate the planets and stars.

It seemed that speculation about space was the only option. But around the time Copernicus proposed the idea that relegated the Earth to one of many celestial objects, technologies were being developed that would bring space within reach for the first time.

2

Rockets and Bombs

Middle Kingdom

Gravity is an implacable adversary in our quest to leave Earth's cradle.

We spend our lives resolutely pinned to the planet. Most people can't jump higher than their waists and even the best high jumpers couldn't clear a one-story building. We do better if we throw something smaller than ourselves. A good athlete throwing a stone upward can probably reach a height of about 70 meters or 230 feet.[1]

Five hundred and fifty years ago, Wan Hu tried to do better. He was a midlevel Ming Dynasty official with an obsession for getting close to the stars. Coming from a rich family, he had a clear path to becoming a high government official, but bureaucracy bored him. Wan Hu was more interested in the Chinese traditions of gunpowder and firecrackers, which had been used for centuries during religious festivals and for entertainment. He also yearned to have a bird's-eye view of the world. Dressed in his finest clothes, Wan Hu sat in a sturdy bamboo chair with forty-seven rockets attached (Figure 4). He held the strings of two kites to guide him on his flight. On his signal, forty-seven assistants lit the fuses and rushed for cover. According to the legend, a tremendous roar

Figure 4. Wan Hu was a legendary Chinese government official of the middle Ming Dynasty (sixteenth century) who tried to become the world's first astronaut by attaching forty-seven rockets to a specially constructed chair.

was followed by billowing clouds of smoke. When the smoke cleared, Wan Hu was gone.

Rather than becoming China's first astronaut, Wan Hu was probably obliterated from the explosive force of so many rockets detonating simultaneously. Despite this spectacular failure, the Middle Kingdom was far ahead of other countries in developing rockets, beginning a long tradition that twinned rocketry with warfare.

The earliest uses of rockets are poorly documented. There are reports that the Greek philosopher Archytas, who speculated about the edge of space, amused the citizens of Tarentum in southern Italy by moving a wooden bird through the air suspended on wires. The propulsion mechanism was escaping steam. The first true rockets were probably accidents. In the first century AD, the Chinese learned how to make simple gunpowder from saltpeter, sulfur, and charcoal dust.[2] They put this mixture into bamboo tubes and tossed them into a fire to make explosions during religious festivals. Some of the tubes may have failed to explode, instead skittering out of the fire propelled by gas from the burning gunpowder.

The word *rocket* appears as early as the third century, during the Three Kingdoms period. Soldiers had learned to attach bamboo tubes filled with gunpowder to arrows, light them, and then launch them with bows. In 228, the Wei State used this kind of "fire arrow" to defend the city of Chencang from the invading forces of the Shu State.

Over the next few centuries, rockets continued to appear in harmless fireworks celebrations, but they showed more promise as military weaponry. The Chinese discovered how to make rockets that could launch themselves. A tube holding gunpowder was capped at one end, with space for a slow-burning fuse, and the other end was left open. The tube was attached to a stick to provide stability and a crude guidance system. When ignited, the outrushing gas from this solid-fuel rocket produced a large forward thrust. Such rockets were first used in battle by the Chinese in 1232 to repel Mongol invaders.[3] Although they weren't effective as weapons of destruction, one can only imagine the psychological effect of being on the receiving end of a barrage of "arrows of flying fire."

Soon, other cultures began their own experimentation and implementation. The Mongols adopted rockets and hired Chinese rocket experts as mercenaries, who helped them conquer Russia and parts of Europe. They used rockets to capture Baghdad in 1258. Quick to learn, the Arabs used rockets ten years later to help defeat Louis IX of France between the Seventh and Eighth Crusades. Europeans soon learned its secrets and started to improve the technology.[4] Roger Bacon discovered the optimum formula for gunpowder: 75 percent saltpeter, 15 percent carbon, and 10 percent sulfur. This recipe was more explosive than Chinese recipes and gave rockets greater range.

Early rockets were so unreliable they could only be used to confuse and frighten the enemy. As the chemistry of gunpowder matured, however, rockets began to influence the outcome of battles. It was the world's first arms race.

The Chinese continued to create new and complex rockets throughout the Ming Dynasty, when even a great seafaring nation like China suffered many thefts and losses due to piracy. To fight the pirates, General Qi Jiguang used hardwood for the body of the rocket and had armor-piercing swords or spears at the front end. He put more than 2,000 rockets on ten warships and developed multishot rockets that allowed up to a hundred of the devices to be launched simultaneously with a single fuse. Other innovations were multistage rockets that could fly for several miles over water and rockets with reusable tubes. General Qi used all of these methods to defend the Great Wall from the Mongols.

The Chinese were first to develop gunpowder and rockets, and during medieval times they were able to keep invaders at bay with formidable arsenals, but their science was based on experimentation with no corresponding development of theory. In the late thirteenth century, mathematician Yang Hui noted: "The men of old changed the name of their methods from problem to problem, so that as no specific explanation was given, there is no way of telling their theoretical origin or basis."[5] Ironically, the stability of Chinese civilization worked against innovation. With a strong central government and stifling bureaucracy, there was little incentive to try something new. Europe, meanwhile, suffered a series of famines and plagues that put an end to growth and caused social upheaval. The Renaissance and the Scientific Revolution emerged from this chaos and propelled Europe to great prosperity.

Ultimately, neglect of science and technology caused the Chinese to lose their edge. By the late fourteenth century, Europe had caught up. For warfare, Europeans developed and perfected the smooth-bore cannon. Rockets were relegated to firework displays.[6] Wan Hu's dreams of traveling to the stars were forgotten.

They were given a firm theoretical basis in the work of Isaac New-

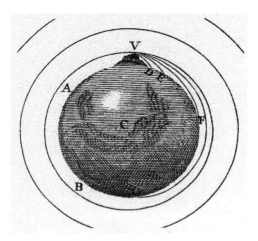

Figure 5. In the thought experiment of Isaac Newton, a cannonball is launched horizontally from a mountain tall enough to be above the Earth's atmosphere. As the velocity increases, the surface curves at the same rate the cannonball falls, creating a circular orbit.

ton, the author of a theory of gravity and laws of motion that would be the basis for space travel centuries later. Newton's 1687 masterwork, *Principia*, unified the terrestrial and celestial realms. Drop an apple and it falls in one second 3,600 times farther than the Moon curves in its orbit, both caused by the action of the Earth's gravity. He described a "thought experiment" where a cannon points sideways at the top of a mountain high enough to be above the atmosphere. With no friction or air resistance, the only force operating is gravity.[7] Fired at modest speed, the cannonball will land at the base of the mountain. As the initial speed is increased, the ball travels farther and farther before landing. Newton calculated the speed where the ball falls toward the Earth's surface at the same rate as the Earth's surface is "falling away" from it (Figure 5).

This is the concept of an orbit. Any projectile shot from Newton's hypothetical cannon at 7.9 kilometers per second or 17,650 mph would remain a captive of the Earth's gravity but would never hit the ground. At over 11 kilometers per second or just over 25,000 mph, the projectile would be liberated from the Earth forever.

The Visionaries

Konstantin Eduardovich Tsiolkovsky was an unlikely rocket scientist. In 1857, he was born into an impoverished family of Polish immigrants in a small Russian town, the fifth of eighteen children. At the age of ten, he developed scarlet fever, leaving him deaf and isolated. By the age of fourteen, his mother had died and he had given up formal schooling.

A reclusive teenager, he moved to Moscow so he could spend long hours at a local library, where he studied physics and astronomy. At the library he was influenced by Nikolai Fyodorov, a futurist who advocated radical life extension and immortality and who thought that the future of humanity lay in space. He also stumbled on the works of Jules Verne and became inspired by Verne's tales of space travel. Tsiolkovsky's family recognized his talent but worried that he was studying obsessively and forgetting to eat. When he was nineteen, his father brought him back home and helped him get a teaching credential so he could earn a living.

Tsiolkovsky became a math teacher in a small provincial school outside Moscow. In his spare time he wrote science fiction, but soon he became more interested in the concrete problems of space travel. He realized that passengers would not survive the acceleration forces of a cannon, the method Jules Verne imagined to get travelers to the Moon. He was far from any center of learning, so when he tried to publish his work on the kinetic theory of gases, a friend had to point out that those ideas had been published twenty-five years earlier. Even as a teenager, Tsiolkovsky had constructed a centrifuge to test the effects of strong gravity. Chickens procured from local farmers were his test subjects. Later, he built the world's first wind tunnel in his apartment and conducted experiments on the aerodynamics of spheres, disks, cylinders, and cones. But he had no funding for his research and he was isolated from the scientific community, so most of his insights were theoretical.

In 1897, however, Tsiolkovsky had an insight that underlies all of space travel today.

He devised an equation relating the change in mass of a rocket to its exhaust velocity. He recognized the critical role of a nozzle in forcing the gas out at high velocity, and he predicted the need for multistage rockets to overcome the Earth's gravity. He also designed fins and gas jets to control the trajectory, pumps to drive fuel into the combustion chamber, and mechanisms that used propellant to cool the rocket in flight. His fertile mind came up with designs for dirigibles, metal jet aircraft, and hovercraft. Hearing about the newly constructed Eiffel Tower triggered the idea of a space elevator as a way of getting into orbit without a rocket.[9]

This Russian visionary continued to face adversity.[10] A year before he developed the rocket equation that bears his name, Tsiolkovsky's son committed suicide. Eight years later, a flood destroyed most of his papers. Three years after that, his daughter was arrested for engaging in revolutionary activities.

In 1911 he wrote: "To place one's feet on the soil of asteroids, to lift a stone from the moon with your hand, to construct moving stations in ether space, to organize inhabited rings around Earth, Moon and Sun, to observe Mars at the distance of several tens of miles, to descend to its satellites or even to its own surface—what could be more insane!"[11] His work took all these ideas from unreal fantasy to the brink of reality.

Tsiolkovsky was sustained in his work by a philosophical and spiritual movement called cosmism. In Russia, one of the foremost proponents of cosmism was Nikolai Fyodorov, whom Tsiolkovsky had met at the library. They shared a utopian belief that the future of humanity was to spread into space and conquer disease and death. Cosmism emerged after the Russian Revolution, envisaging a heroic image of the proletarian who strides forth from the Earth to conquer planets and stars.[12] One quote epitomizes Tsiolkovsky's views on space: "The Earth is the cradle of humanity, but mankind cannot stay in the cradle forever."

In the 1920s, the young physicist Hermann Oberth was unaware of Tsiolkovsky's work, but he too dreamt of space travel. Like the Russian, Oberth was inspired by Jules Verne, rereading the novels to the point of memorization. He dabbled with rockets as a child and by 1917 his expertise had grown such that he fired a rocket with liquid propellant as a demonstration for the Prussian minister of war.[13] His doctoral thesis, "The Rockets to the Planets in Space," later became an essential contribution to rocket science, but initially it was rejected. Oberth was fiercely critical of the German education system, saying it was ". . . like an automobile which has strong rear lights, brightly illuminating the past. But looking forward, things are barely discernible."[14]

Like Tsiolkovsky, Oberth worked outside academia for the majority of his career, earning a living as a schoolteacher. He was a leading member of the "Spaceflight Society," a German amateur rocketry group whose members scavenged any materials they could find for their rockets as Europe descended into an economic depression. In 1929, Oberth was a technical adviser to the film pioneer Fritz Lang for *Woman in the Moon*, the first film ever to have scenes set in space. He lost an eye during a publicity stunt for the film. That same year, he conducted a captive firing of his first liquid-fueled rocket engine. One of his assistants was eighteen-year-old Wernher von Braun, who would later feature prominently in our efforts to reach space.

The first to launch a liquid-fuel rocket was American Robert Goddard. As a boy, Goddard was thin, frail, and subject to pleurisy, bronchitis, and stomach problems. He spent much of his time holed up in the local public library, where he was transported by the science fiction of H. G. Wells. Goddard fixed his inspiration to a day when he was seventeen and he climbed a cherry tree to remove dead limbs: "I imagined how wonderful it would be to make some device which had even the *possibility* of ascending to Mars, and how it would look on a small scale, if sent up from the meadow at my feet. . . . I was a different boy when I descended the tree from when I ascended."[15]

In 1914, Goddard registered the patents for a liquid-fuel rocket and a multistage rocket, the first of his more than two hundred patents. He was a hands-on experimenter as well as an expert physicist. Liquid-fuel rockets are finicky because the volatile fuel and oxidizer must be injected into a combustion chamber at a carefully controlled rate. On a bitterly cold spring morning in 1926, Goddard achieved success with a small liquid-propellant rocket dubbed "Nell." Launched from his Aunt Effie's farm, it traveled for 184 feet in a flight that lasted less than three seconds, landing in a cabbage field (Figure 6). Over the years, he con-

Figure 6. Robert Goddard is bundled against the cold of a New England winter in 1926 as he stands by the launching frame of his most notable invention. The liquid fuel of this rocket was gasoline and liquid oxygen, contained in the cylinder across from Goddard's torso.

ducted more than three dozen test flights, refining his designs and techniques until he reached altitudes of several miles. In 1929, he began
what became a lifelong friendship with Charles Lindbergh, who shared
his vision.[16]

Nevertheless, the world was not quite ready for rockets. Goddard's
seminal paper from 1919, "A Method of Reaching Extreme Altitudes,"
was ridiculed by the press and fellow scientists. An unsigned editorial in
the *New York Times* was particularly harsh, accusing him of ignorance
of the laws of physics: ". . . Professor Goddard . . . does not know the
relation of action and reaction, and of the need to have something better than a vacuum against which to react. . . . Of course he only seems
to lack the knowledge ladled out daily in high schools."[17] Forty-nine
years after ripping Goddard, and a day after the launch of Apollo 11, the
paper issued a brief correction: "Further investigation and experimentation have confirmed the findings of Isaac Newton in the 17th Century
and it is now definitely established that a rocket can function in a vacuum as well as in an atmosphere. The Times regrets the error."[18] The
apology was too late for Goddard, who died of throat cancer in 1945.

Wernher von Braun

Warfare and space exploration merged again in the 1940s. Goddard had
financed his research with small grants from the Smithsonian Institution and the Guggenheim Foundation; no government agency showed
interest and the military was particularly dismissive. But America's
future adversaries were very interested in Goddard's rocketry. During
the 1930s, a German military attaché working in the United States sent
a report on Goddard's work back to the military intelligence agency,
and the Soviets gleaned information from a KGB spy embedded in the
US Navy Bureau of Aeronautics. Toward the end of World War II, Goddard got to inspect a captured German V-2 ballistic missile. The V-2

was far more advanced than any of Goddard's rockets, but he was convinced the Germans had "stolen" his ideas. In particular, Goddard was furious at Oberth, whom he accused of plagiarizing his 1919 work; this episode contributed to Goddard's secrecy and paranoia.[19]

The architect of the V-2 was the most controversial figure in the history of rocketry: Wernher von Braun.

We can picture the young German boy as he became hooked on rockets. Inspired by Germans who were setting land speed records in rocket-propelled cars, the twelve-year-old caused major disruption in a crowded street. Echoing Wan Hu, he attached to a toy wagon a dozen of the largest skyrockets he could find. Rather than riding the wagon as Wan Hu had ridden his sedan chair, von Braun lit the fuses and stood back. He was thrilled with the results: "It performed beyond my wildest dreams. The wagon careened crazily about, trailing a tail of fire like a comet. When the rockets burned out, ending their sparkling performance with a magnificent thunderclap, the wagon rolled majestically to a halt."[20] The police who arrived on the scene were less impressed; they took the young boy into custody.

Wernher von Braun was rescued from that indiscretion by his father, who was the German minister of agriculture. His mother could trace her ancestry back to the kings of France, England, and Denmark, and the young von Braun inherited the title of baron. All through his life, he exhibited a self-confidence bordering on arrogance.

Though he was a gifted musician who played piano and cello and composed in the style of Hindemith, von Braun initially struggled with math and physics. His mother bought him a telescope, allowing him to be captivated by the Moon. As a young teenager, he bought *By Rocket into Interplanetary Space* by Hermann Oberth but was dismayed when he opened it. He recalled, "To my consternation, I couldn't understand a word. Its pages were a baffling conglomeration of mathematical symbols and formulas."[21] He realized that the success of space travel was underpinned by technical calculations, so he decided to master the relevant subjects. At the age of eighteen, he began his long tutelage with Oberth; that same year, he attended a talk by a pioneer of high-altitude ballooning, telling him, "You know, I plan on traveling to the Moon at some time."

When Adolf Hitler came to power, Wernher von Braun was twenty-one. He later claimed that he had been apolitical and disinterested in the world around him. But his uncritical patriotism meant that, at best, he was surprisingly naïve about the ramifications of his work and, at worst, he was complicit in death and destruction.[22]

With the joy of an amateur, von Braun continued to experiment with rockets. His days in Berlin were busy with study toward a graduate degree in physics, but he spent every spare moment at a derelict, 300-acre site of scrub and weeds at the edge of the city. There, members of the Berlin Rocket Society carried out their work using scrounged materials and donated labor. When the Army Ordnance Department took interest and started to fund their research, von Braun was delighted. (It was in fact exactly the type of military support that Goddard had sought and failed to get for his own work.) When von Braun finished his thesis in 1934, parts of it were considered so crucial to national security that they stayed classified until 1960. He set aside his dreams of space travel and moved to a big facility that the Army built for him on an island in the Baltic Sea. There he worked on a weapon the Nazi Propaganda Ministry would eventually call the Vengeance Weapon 2, or the V-2 (Figure 7).

Even if science, not politics, motivated von Braun, he was part of the machinery of war. He joined the Nazi Party and the SS, and photos exist of him donning those uniforms and posing in the company of senior Nazi Party members. After seeing film footage of the successful launch of a V-2 prototype, Hitler personally made von Braun a professor—an exceptional honor for a thirty-one-year-old engineer.

The V-2 was inaccurate but effective as a "terror" weapon, a projectile screaming out of the sky at four times the speed of sound, impos-

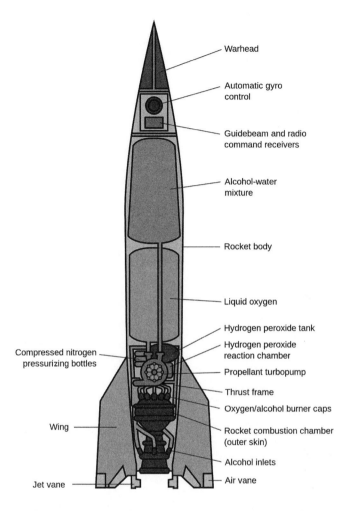

Figure 7. Schematic diagram of the German A4 rocket, later renamed the V-2, or Retaliation/
Vengeance Weapon 2. It was the world's first long-range ballistic missile and more than 2,000
of them were launched toward England and Belgium in the latter part of World War II.

sible to defend against. The rockets killed an estimated 9,000 civilians
and military personnel in London and Antwerp during Nazi airstrikes.
Each rocket was made in an underground factory at Mittelwerk, where
prisoners from the nearby Dora-Mittelbau concentration camp toiled in
deplorable conditions. About 12,000 forced laborers and prisoners died
producing the weapons.

But von Braun's insider status didn't place him above the Nazi Party's suspicion. At an event in early 1944, after drinking too much, he said he thought the war would end badly for the Germans and that all he'd ever wanted to do with his rockets was send them into space. Such talk was tantamount to treason. He was a pilot, so the Gestapo arrested him to keep him from defecting to the West. A month later, Albert Speer convinced Hitler to release von Braun because he was critical to the V-2 program.

In early 1945, as Allied forces moved deep into Germany, the SS moved von Braun and his team to the Bavarian Alps, with orders to execute them rather than let them fall into enemy hands. But von Braun argued for the team to be dispersed so as not to be an easy target for American bombers. He had heard stories of the harsh treatment meted out by the Soviets to their captured enemies, so he deliberately surrendered to the Americans rather than the approaching Soviet forces. On May 2, he was able to slip away and surrender to a private from the 44th Infantry Division. He was at the top of the blacklist of German scientists and engineers targeted for interrogation by US military experts.[23]

When the fog of war lifted, von Braun had been rehabilitated. The American intelligence agencies created a false employment history for him, expunged his Nazi Party membership from the public record, and gave him a security clearance. Although he should have been relieved to get through the war unscathed, von Braun chafed at the restrictions of his new life. Working at Fort Bliss near El Paso, Texas, he couldn't leave the base without a military escort. Whereas in Germany he had had thousands of engineers reporting to him when he was only twenty-six, in the United States he had a small team and was starved of resources. At least his loyal German engineers continued to address him as Herr Professor.

Although the postwar years were frustrating for von Braun, he gained a new start and had been "cleansed" of his Nazism. He was free to pursue his dreams of space.

The Big Chill

Germany lost the war due to a "marriage of convenience" between the Soviet Union and the Western allies. But those countries' ideological differences bubbled up in the aftermath of the war, setting the stage for the *Cold War*, a term coined by writer George Orwell in October 1945.

As the war ended, Wernher von Braun and a hundred senior German scientists were working under US Army command with orders to continue development of the V-2 rocket.

Meanwhile, the Soviets took over jurisdiction of the Mittelwerk factory but found that most of the best engineers had already defected to the Americans. Whereas in the United States the Germans were at the core of rocket development, the Germans who worked in the Soviet Union were used only as consultants and were repatriated in the early 1950s. The Soviet counterpart to von Braun was the equally brilliant Sergei Korolev. He started by reverse engineering the V-2 but quickly developed his own designs, leading to a 100-ton engine of unprecedented power. As a result of one of Stalin's purges, Korolev spent six years in prison, where mistreatment led to serious health problems throughout his life. The Soviets referred to him only as the "Chief Designer" during the Cold War and his identity wasn't revealed in the West until after his death in 1966.

Mistrust between the United States and the Soviet Union deepened after the war. The United States lost its monopoly on the atomic bomb and watched helplessly as the Soviets annexed European countries to form an "iron curtain" that stretched from the Baltic to the Adriatic. The Soviets had suffered 27 million casualties in the war and they feared invasion, especially as the United States had a far superior air force with bases near Soviet territory. The role of ideology in the so-called Space Race has been summarized by journalist and historian William Burrows: "The cold war would become the great engine—the supreme catalyst—that sent rockets and their cargoes far above Earth and worlds away. If Tsiolkovsky, Oberth, Goddard, and others were the fathers of rocketry, then the competition between capitalism and communism was its midwife."[24]

The visionaries never gave up their ambitions for humans to leave the Earth. But for the next ten years, dreams of space travel were overshadowed by nightmares of nuclear holocaust.

The United States didn't clash with the Soviet Union directly. The rivalry played out as a toxic brew of military jockeying, proxy wars, support of strategic allies, espionage, propaganda, and technological and economic competition. The cutting edge of the Cold War was a nuclear arms race. As the war ended, America was confident of its advantage in developing nuclear weapons, so American experts were shocked when the Soviets exploded their first atomic bomb in 1949.

The Manhattan Project was the US research and development project to build an atomic bomb. It had been so secret that Vice President Harry Truman was unaware of its existence, but it was riddled with spies. Both countries began massive investment in their arsenals. America was the first to detonate a hydrogen bomb in 1952, but the Soviets followed suit less than a year later.

The Space Race began when the United States and the Soviet Union developed ballistic missiles that could launch objects into space. It was kicked off in 1955 by announcements only four days apart that both nations were planning to launch artificial Earth satellites.[25] Stockpiles of nuclear weapons grew rapidly; the underlying goal was the capability to target and destroy any city in the opposing country within hours.

In the United States, development of intercontinental ballistic missiles was hampered by the three branches of the military each wanting their own capabilities. The Air Force developed the Atlas rocket, the Navy had the Vanguard rocket, and von Braun's US Army team was perfecting the Redstone rocket, a direct descendant of the V-2. As the Space Race intensified, President Dwight Eisenhower gave the nod to the Navy, since its Vanguard rocket was being developed by the Naval Research lab, which was seen as a scientific rather than a military organization. The Atlas and Redstone programs were iced. Eisenhower wanted to avoid the overt militarization of space, and he didn't want to hand the Soviets a propaganda victory.

Meanwhile, the Soviets pursued intercontinental ballistic missiles with a relentless focus and ample funding. Sergei Korolev developed the R-7, which was more powerful than any American rocket; it could deliver a three-ton warhead 5,000 miles. Derivatives of the R-7 have been used in the Soviet and the Russian space programs for more than fifty years. On October 4, 1957, the Soviets shocked the world when they launched into orbit a beeping metal sphere the size of a beach ball and the weight of an adult man: the satellite Sputnik 1 (Figure 8).

Stakes in the Space Race suddenly became very high. The Americans and the Soviets each developed satellites as part of the International Geophysical Year of 1957–1958, which, ironically, was a project conceived after Stalin's death to thaw the Cold War. While small satel-

Figure 8. Sputnik 1 was the first artificial Earth satellite. Launched by the Soviet Union into a ninety-minute, low Earth orbit in October 1957, it transmitted signals for twenty-two days before burning up in the atmosphere. Sputnik triggered the Space Race.

lites could be used for scientific research, the deeper concern was that large satellites could put nuclear weapons into orbit. Whoever controlled the frontier of space would control the world.

The United States rushed to match the feat of Sputnik, but the country experienced humiliation when a live national TV audience watched a Vanguard rocket explode seconds after launch. Newspapers called it "Flopnik" and "Kaputnik," and the Soviet UN delegate offered the United States aid under a "Soviet program of assistance to backward nations." Called into action, von Braun and his team rose to the challenge, and Explorer 1 was launched into orbit on January 31, 1958. It was a face-saving success, but the Soviets still held an edge: Sputnik weighed 84 kilograms, as much as a grown man, while Explorer weighed in at just 5 kilograms, not much heavier than a brick.

Although it was the height of the Cold War, and space could easily have become the exclusive preserve of the military, the cool head of the commander-in-chief prevailed. Eisenhower was playing catch-up with a well-funded and better-organized adversary. As a former general, Eisenhower knew enough about military bureaucracy to prefer a civilian organization. He also knew that the best innovation would come from a national space agency rather than small groups often working in competition and in isolation. When Congress held hearings on the subject, one of the key players forcing the pace was a young senator from Texas, Lyndon Johnson.[26]

The events leading to the formation of the National Aeronautics and Space Administration (NASA) laid bare the tension among the competing motivations to go into space. As Sputnik orbited the Earth, Eisenhower appointed James Killian as his special assistant for science and technology. Killian was the president of MIT, so his selection provided a strong signal that Eisenhower wanted to keep a civilian emphasis on space. Late in 1957, Killian wrote a memo to Eisenhower saying that many scientists strongly opposed Defense Department control of the space program because it would limit space research to military objectives and tar all US space activity as military in nature. Meanwhile, the Senate Armed Services Preparedness Subcommittee heard from dozens of experts who said the United States would never have been beaten to the punch by Sputnik if the military had held sway. Then in May 1958, the Soviets launched the one-ton Sputnik 3, and the sheer size of the satellite triggered recriminations and new calls to action. The voices of the hawks grew louder.[27]

Eisenhower held firm. On October 1, 1958, he established NASA as a civilian organization for the peaceful exploration of space. The new agency began with 8,200 employees and a budget of $340 million. Its charter in the Space Act had eight objectives, which included expanding knowledge of space, improving space vehicles, preserving the leadership of the United States in space science and technology, and collaborating with our international partners and allies.[28] The Space Act was signed a little less than a year after the beeping Soviet satellite had rocked the world.

3

Send In the Robots

Fly Me to the Moon

Sputnik was a technological "Pearl Harbor moment" for America, but the feeling of scrambling to keep up with the Soviets continued.

In the first five years of their space program, the Soviets reeled off an impressive series of achievements: first satellite in orbit, first object to leave Earth's gravity, first data link to space, first probe to crash-land on the Moon, first probe sent to Venus, first probe sent to Mars, first man in space, first woman in space, first dual manned spaceflights, and, last but not least, the first dogs to be put into orbit and safely returned to Earth.[1]

Yuri Gagarin became the first human to orbit the Earth on April 12, 1961, rising from the barren steppes of Kazakhstan aboard the Vostok 1 spacecraft. His modest stature—Gagarin was just 5 feet 2 inches—helped him fit into the tiny capsule. His flight lasted only one orbit and Gagarin didn't control the spacecraft; it was flown in automatic mode as a precaution, since medical science didn't know what would happen to a human subjected to the stresses of launch and subsequent weightlessness. Gagarin had the ability to take over control of the spacecraft in an

Figure 9. Yuri Gagarin became the first human in space on April 12, 1961. He retired with the air force rank of colonel and received the award of Hero of the Soviet Union, the nation's highest honor. Gagarin was a worldwide celebrity; he died in a routine training flight in 1968.

emergency by opening an envelope and typing a special code into the computer.[2] Nevertheless, Vostok 1 was hailed as a historic event (Figure 9).[3] The United States experienced an echo of the shock and embarrassment felt with the launch of Sputnik.

A new young president responded quickly. Less than two months after Gagarin's flight, and less than three weeks after Alan Shepard became America's first astronaut in a fifteen-minute suborbital flight, John F. Kennedy addressed a special joint session of Congress: "I believe that this nation should commit itself to achieving the goal, before this decade is out, of landing a man on the Moon and returning him safely."[4]

The first phase of the manned space program took place against a backdrop of Cold War escalation. Even though the first astronauts in the Mercury program didn't fly their spacecraft, putting humans in orbit was concomitant with the larger goal of mastery of space. Mastery of space in turn was seen as a vital tool in the power struggle between the two superpowers. After the failure of Kennedy's covert plan to topple Fidel Castro, the Soviets increased their military support for Cuba. In Europe, American and Soviet tanks were facing off across the newly constructed Berlin Wall. When the Soviets prepared to install nuclear missiles in Cuba in October 1962, it felt as if the world had come to the brink. The United States had more than 30,000 nuclear weapons and the Soviet Union was rapidly catching up. The deterrence logic of "mutually assured destruction" was scant solace.

So began the Apollo program, the largest and most complex technical undertaking in human history.[5] At its peak, it involved 500,000 people and 20,000 companies. Its cost in present-day dollars was more than $100 billion.

To get to the Moon so quickly, NASA needed a huge budget and a tight and single-minded focus on the goal. At the time of Kennedy's speech, only two humans had ever traveled in space. As a precursor to Apollo, NASA began Project Gemini in 1962. Gemini spacecraft carried two astronauts and the missions tested docking technology, practiced working outside a spacecraft, and orbited long enough to mimic a trip to the Moon and back. All the early Apollo crews were veterans of the Mercury and Gemini programs.

America was hopeful that the two superpowers might collaborate rather than duplicate the vast effort required for a race to the Moon. After stepping back from the brink following the Cuban missile crisis, Kennedy and Soviet Premier Nikita Khrushchev had developed a mutual understanding. In 1963, in a speech to the United Nations General Assembly, Kennedy proposed a joint space effort. Khrushchev initially rejected the overture but was poised to accept it when Kennedy was assassinated in November 1963. Even Kennedy was hedging his bets, though, wavering between cooperation and competition. In the speech he never lived to give on November 22, he was to have said, "The United States of America has no intention of finishing second in space."[6] Within a year, Khrushchev had been ousted and Lyndon Johnson and Leonid Brezhnev had a frosty relationship, so joint ventures in space were taken off the table.

The relationship between the world's two superpowers was plagued with misinformation and misperceptions. This has only become clear in hindsight and with the release of documents that were classified at the time.[7] Each country was fearful of the other, and each overestimated the capabilities of the other. The Soviets refused to cooperate in space in part because they didn't want to expose the technical shortcomings of their program. With his famous 1961 speech, Kennedy in effect dared them to show their stuff.

Space travel in the early days was risky. Most of the Soviet losses were kept secret at the time but have come to light since.[8] In 1960, more than a hundred top Soviet military and technical personnel were killed when the second stage engine of an R-16 rocket ignited the propellant tanks of the first stage, causing an explosion and fire. Chief Marshal of Artillery Mitrofan Nedelin was vaporized, and his only identifiable remains were his war medals. A year later, a Soviet cosmonaut was incinerated in a fire in a test chamber with high oxygen content. The Soviets erased all evidence that he had ever existed, which was particularly tragic, since the Apollo 1 flight crew died in similar circumstances in 1967; knowing about the Soviet incident might have led to a redesign of the capsule. As it happened, a spark ignited an oxygen fire in their capsule during a ground test, and Gus Grissom, Ed White, and Roger Chaffee were severely burned and then asphyxiated after all the oxygen had burned away. The same year, Vladimir Komarov died after a trouble-plagued flight in Soyuz 1, when the parachute system failed to slow his descent.

The men who strapped themselves into a small metal container on top of half a million gallons of kerosene and liquid oxygen were extraordinarily brave. People who witnessed a Saturn V launch recounted that even at a distance of two miles, its engines produced coruscating heat and waves of pressure that passed through the ribcage. The five massive engines gulped 15 tons of fuel each second and produced eight million pounds of thrust. The giant rocket was 60 feet taller than the Statue of Liberty (Figure 10). In July 1969, there were plenty of white knuckles at Mission Control in Houston when Neil Armstrong assumed man-

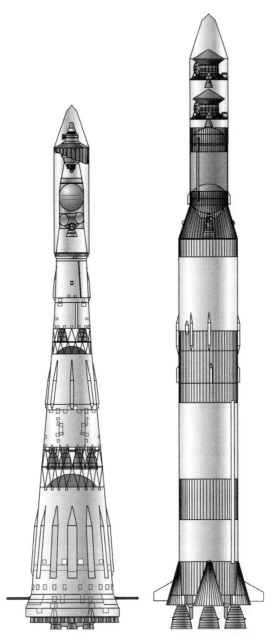

Figure 10. A comparison of the Soviet N1/L3 rocket (left) and the US Saturn V rocket (right). The Saturn V was as tall as a thirty-six story building; it had a peak thrust of eight million pounds of force and could lift 60 tons to Earth orbit.

ual control of the Apollo 11 landing module, after a series of technical glitches, and guided it over a field of rugged boulders to a soft landing on the Moon with less than a minute of fuel left. Back in Houston, Charles Duke radioed him: "You got a bunch of guys about to turn blue. We're breathing again."[9]

The Moon landings were and still are unprecedented. The twenty-four men who journeyed there are the only people ever to have left the Earth's gravity, and the twelve who landed are the only people to have set foot on another world.

Despite this feat, and the ingenuity and heroism of the Apollo 13 crew, who in 1970 nursed their crippled spacecraft back to Earth following an oxygen explosion, public interest in the Moon landings waned. Through a misty lens of history, it seems the Apollo program had broad public support. But in fact a majority thought the government was spending too much on space. Both Kennedy and Johnson complained about the enormous cost of the Apollo program, and the final three planned Moon landings were canceled to allow NASA to start work on the Space Shuttle, which was intended to be a "space truck" that could routinely haul astronauts and cargo into low Earth orbit. In effect, it was a retreat from the grandiosity of the Apollo missions.

Yet something profound happened as a result of the Moon landings.

The astronauts were patriots, but they instinctively knew they were representing all of humanity. As they orbited the Earth, many commented on the seamlessness of a planet where no political or cultural boundaries were visible. The iconic image of the fragile Earth hanging in the blackness of space—a blue marble—helped spur the environmental movement in the late 1960s. It is indeed ironic that a supreme feat of the military-industrial complex was embraced by counterculture activists.[10] When Frank Sinatra performed "Fly Me to the Moon" on his TV show in 1969, he dedicated it to the astronauts who had "made the impossible possible." The song's jaunty melody perfectly captured the lightness of the people who had slipped the bonds of Earth.

Of Mice and Men

The hardest part of space travel is getting there.

For the rocket, the key quantity is Max-Q—the maximum aerodynamic stress due to drag from the atmosphere as the rocket accelerates. The stress is small at low altitude because the speed is lower, and small at high altitude because the atmosphere is thin. Somewhere in between is Max-Q, the moment when engineers watching a launch hold their breaths. For both the Saturn V and the Space Shuttle, Max-Q occurred about a minute after launch, at an altitude of about 40,000 feet.

For any occupant of the rocket, there's buffeting and vibration, but the maximum hazard is presented by g-forces. We spend our lives subject to a downward acceleration of 9.8 meters per second per second, or 1 g. As flexible, water-filled sacks, we're fairly tolerant of acceleration, but it depends on the direction. Fighter aircraft pilots can handle a positive 8 or 9 g's, where the blood is being forced to the feet, as long as it lasts no more than a few seconds. But minus 2 or 3 g's, when blood is being forced into the head, can cause blackouts and even death. Air Force Colonel John Stapp, a flight surgeon, risked his life to test these limits in the 1950s. Stapp was repeatedly strapped into a rocket sled, and in one test he survived a momentary force forty-six times stronger than normal gravity. The colonel suffered broken limbs and permanent vision loss due to these experiments, but he still managed to die peacefully at home at the age of eighty-nine.

The Apollo astronauts felt a maximum of 4 g's just before the huge main-stage engines shut off, and close to 7 g's when they reentered the Earth's atmosphere. Space Shuttle astronauts, on the other hand, pulled no more than 3 g's on either ascent or descent, something you could experience on any decent roller coaster. But early in the Space Age, medical science was unsure if people could survive the rigors of space, so a lot of experiments were done using mammals as test cases.

This continued a long tradition; in 1783, a sheep, a duck, and a rooster were sent up in the recently invented hot-air balloon.

Laika is one of the unsung heroes of spacefaring. She was a husky-terrier mix, a stray dog found wandering the streets of Moscow. Soviet scientists preferred strays because they thought life on the streets would have made them resilient. Laika was chosen from among ten dogs due to her phlegmatic temperament. After being subjected to centrifuges and noisy environments, she was conditioned for the tiny capsule by being confined in successively smaller spaces for periods of up to three weeks. Nikita Khrushchev put great pressure on mission designers, wanting a launch in time for the fortieth anniversary of the Bolshevik Revolution. So Sputnik 2 was prepared in a hurry and launched, with Laika aboard, less than a month after Sputnik 1.

Early data showed that Laika was agitated but eating her food. However, the temperature-control systems were inadequate, and she died from overheating and stress after seven hours in orbit. There was never any possibility of her surviving the flight; poisoned food had been prepared to euthanize her before the fiery reentry. At the time, it was reported that she died when her oxygen ran out on the sixth day of the flight. Animal rights groups protested at Soviet embassies around the world, and there was a demonstration at the United Nations in New York.[11] Years later, when the Soviet Union fell and scientists could speak freely, some did express remorse. Laika's trainer, Lieutenant General Oleg Gazenko, admitted, "Work with animals is a source of suffering to all of us. We treat them like babies who cannot speak. The more time passes, the more I'm sorry about it. We shouldn't have done it. . . . We did not learn enough from this mission to justify the death of the dog."[12]

While the Soviets used dogs, the Americans preferred monkeys due to their similarity to humans. The first monkey in space was Albert, launched on a V-2 rocket in 1948. Albert died of suffocation. For the first decade of such experiments, the fatality rate was very high. In 1959, Able and Baker became the first US animals to fly into space and return

Figure 11. "Miss Baker," a female squirrel monkey from Peru, was the first monkey to survive spaceflight. She ascended to an altitude of 360 miles in the nose cone of a US Air Force ballistic missile, surviving 32 g's and reaching a top speed of 10,000 mph.

alive, withstanding 32 g's along the way. Able was a rhesus monkey who died soon afterward during a surgical procedure, but "Miss Baker," a squirrel monkey, survived another twenty-eight years (Figure 11). She got as many as 150 letters a day from children and was buried on the grounds of the US Space and Rocket Center in Huntsville, Alabama. Three hundred people attended her funeral.

Fruit flies were the first animals of any kind sent into space, aboard a captured Nazi V-2 rocket in 1947. They were followed by mice, then monkeys, then men and women.

Since then, a menagerie of animals has made the trip. By the early 1960s, both the Americans and the Soviets had launched mice into space, and the Soviets added frogs and guinea pigs to the launch personnel. France got into the act with rats, and in 1963 they planned to launch Felix the cat, but Felix had other plans and he escaped, so they sent up Félicette instead. In 1968, two tortoises became the first animals to go to the Moon, aboard Zond 5. They were accompanied by wine flies, mealworms, and other biological specimens. A few years later, America sent mice and nematodes to the Moon on Apollo 16 and Apollo 17. The Space Shuttle facilitated animal space travel, and now spiders, bees, ants, silkworms, butterflies, newts, sea urchins, and jellyfish have all been in orbit. Astronauts have had reason to be wary of some of their passengers, especially Madagascar hissing cockroaches and South African rock scorpions.

Most of these hazardous trips were to low Earth orbit, a few hundred miles, or just an afternoon's drive straight up. Even the round trip to the Moon is less than half a million miles, a distance many business people rack up every few years on atmospheric jet travel.

By contrast, the planets seem far beyond reach.

Exploring the Planets

NASA's budget never again reached the giddy heights of the mid-1960s. As a percentage of the federal budget, NASA soared from its inception to a peak of 5.5 percent in 1967 and then fell just as rapidly down to 1 percent in 1973. It has bumbled along below 1 percent ever since.[13] In the 1970s, the agency embraced a different challenge, albeit not one as grand and dramatic as having astronauts cavort and drive on the Moon.

A critical transition in the history of ideas is the shift from an Earth-centric worldview, where our planet is seen as special and unique, to a "many worlds" concept, where objects in space are physically and geologically familiar. Space travel brings those worlds into view in a way that can't be approached by telescopic observation.

Before 1610, the planets were just nontwinkling dots that drifted across a celestial backdrop. The Moon had craters and dark "seas" that the eye could interpret as imaginary figures. When Galileo pointed a telescope at the Moon, he observed a surface that, ". . . just like the face of the Earth itself, is everywhere full of vast protuberances, deep chasms, and sinuosities."[14] But this paled when compared with what we learned when Apollo astronauts went there, walked over the rugged terrain, and returned with 842 pounds of rocks. Now we know the Moon's age to within an accuracy of a percent, we know its geological history, and we know it formed out of debris from an impact on the infant Earth.

Several hundred years of observations with telescopes uncovered a handful of additional planets but revealed almost nothing about their true nature. They remained small, blurry disks of light. The exception was Mars, which had pale poles and a network of features that, in the wishful thinking of amateur astronomer Percival Lowell, represented an irrigation system of a Martian civilization. Even a nearby planet like Mars is so far away that telescopes reveal little of its physical reality. As recently as 1966, scientists still argued over whether or not Mars was covered with vegetation.

The context for understanding planetary exploration is the vastness of space. When we progressed from orbiting the Earth to landing on the Moon, it was like leaving our backyard to explore another city. Earth orbit is a few hundred miles up, while the Moon is a quarter of a million miles away. That increase of a factor of a thousand severely taxed our ingenuity. Compared to the Earth–Moon distance, the distance to Mars at its closest approach is 200 times greater, and the distance to Jupiter at its closest approach is 1,600 times greater. Jumping to the edge of the Solar System is another factor of a thousand.

While sending men to the Moon was the Space Race's big prize, the Americans and the Soviets could test their technology and expand their knowledge of the Solar System by guiding robotic spacecraft to targets hundreds of millions of miles from Earth. Failures were common. In 1958, the Army and the Air Force saw four failed launches of the Pioneer series of probes. Meanwhile, the first three launches of the Luna program also failed, and the Soviets got in the habit of not disclosing launches that failed to reach orbit and not even assigning them a Luna number. But with persistence came success. In January 1959, less than two years after Sputnik shocked the world, Luna 1 was the first man-made object to leave Earth's gravity. By the end of 1959, its successors Luna 2 and Luna 3 had crashed into the Moon's surface and taken photos of its crater-pitted dark side. The scientific payoff from these probes was substantial, yielding information on the chemical composition, gravity, and radiation environment of the Moon.

In 1962, the United States achieved the first planetary flyby, as Mariner 2 swooped within 20,000 miles of Venus. Two years later, Mariner 4 executed the first flyby of Mars. Getting a planetary probe to its target was a technical tour de force. In a golfing analogy, the flybys were like hitting the ball 400 yards off the tee to within an inch of the hole.

Landing spacecraft and returning data from these planets was much harder.

The Soviets succeeded first, when Venera 7 landed on Venus and sent back twenty-three minutes of data in 1970. But that was after fifteen attempts. Before the Venera series, three spacecraft failed to leave Earth orbit and another exploded. When the Soviet Mars 3 lander sent back less than twenty seconds of data in 1971, it came after seven failed missions. The Soviets had such trouble with their Mars missions that they gave up trying to get there for more than a decade.

As NASA engineers started to apply their expertise to exploration of the Solar System, the agency realized they had no one there to do the science. So they cajoled and bribed universities into hiring faculty and postdocs, and the academic field of planetary science was born. The field was an amalgam of geology and astronomy and attracted its fair share of iconoclasts and larger-than-life figures.

The learning curve for planetary exploration was brutal, but the results were spectacular. In the wake of Apollo, a young and ambitious cadre of planetary scientists worked with NASA to launch twin orbiters and landers to Mars (Viking 1 and Viking 2), probes to Jupiter and

Saturn (Pioneer 10 and Pioneer 11), and probes to outer planets Uranus and Neptune (Voyager 1 and Voyager 2). These missions from the 1970s were hugely successful. The Pioneer 10 and 11 probes, launched in 1972 and 1973, respectively, both flew by Jupiter and its moons; Pioneer 11 got a close look at Saturn too. They each carried a golden plaque etched with human figures and information about the origin of the probes, in case aliens would one day find them. Both have left the Solar System, and Pioneer 10 is more than 10 billion miles from home. The Voyager spacecraft are both still transmitting data more than thirty-six years after their launch in 1977. Voyager 2 visited Uranus and Neptune and Voyager 1 is the most distant human artifact, coasting through interstellar space 12 billion miles from Earth. The Viking spacecraft launched in 1975 released twin landers to different locations on Mars, where they conducted the first and only tests of life in the Martian soil. This was indeed the "golden age" of planetary science (Figure 12).

The pace slowed and planetary science was in the doldrums in the 1980s, but Cassini was launched in 1997 and is still exploring the Saturn system. Cassini is the size of a bus and bristles with a dozen scientific instruments. The spacecraft that traveled a billion miles to view new worlds for the first time saw amazing sights: fractured ice on top of an ocean on the water world Europa, lakes of ethane and methane on

Figure 12. The first image ever returned from the surface of another planet. Viking 1 landed on Mars on July 20, 1976. The close-up view of Mars as an arid, frigid desert replaced decades of speculation about the red planet. The rock near the center is 10 cm across.

Titan, volcanoes on Io coating the tiny moon with an inch of sulfur every year, moons as dark as soot and as bright as a mirror. Cassini dropped the Huygens probe for a soft landing on Titan in 2005, revealing an exotic world with lakes and rivers, clouds and rain. Huygens was a 300-kilogram spacecraft that sampled the atmosphere and took pictures of the surface for a few hours before its battery died. It's still the most distant landing of any manmade craft.[15]

Digital cameras on planetary probes showed that these remote worlds had distinctive features and "personalities." Instead of a single pixel, the new cameras transmitted millions of pixels. In 1990, when Voyager 1 got to the edge of the Solar System after traveling for twelve years and four billion miles, it reversed the usual situation. Looking back, it snapped a picture of the Earth, which was no more than a mote of light on a dark backdrop. Carl Sagan called this evocative image the "Pale Blue Dot," and he used it to issue a clarion call for humanity to get its house in order: "Our planet is a lonely speck in the great enveloping cosmic dark. . . . there is no hint that help will come from elsewhere to save us from ourselves. . . . This distant image of our tiny world . . . underscores our responsibility to deal more kindly with one another, and to preserve and cherish the pale blue dot, the only home we've ever known."[16]

Man versus Machine

The difficulties we have getting there remind us of the fact that we weren't made to live in space. Let's consider what happens to an unprotected human there.

Imagine you're at one end of a large space station in a room with air but no food or water. You have no spacesuit. Safety lies at the end of a long tunnel whose wall has been ruptured by a meteorite impact, leaving pure vacuum inside. You estimate it will take you five seconds to propel yourself to the end of the tunnel and perhaps another ten to open the air lock and get into a pressurized zone. Would you make it?

Not if you took a deep breath. Vacuum is lethal because it makes the air in your lungs expand, rupturing delicate tissue, so emptying your lungs would be a better strategy. Water in your tissues would vaporize and bubbles would form in your veins, but your skin would likely stop you from exploding. It would be unlikely, however, that you could get to safety before lack of oxygen flow to the brain caused you to pass out, which takes roughly fifteen seconds. Death follows in a minute. Compared with these indignities, adjusting to zero gravity is a walk in the park.

Humans have proved capable of living and working in space, but their fragility and the cost of keeping them safe has spurred a long debate over whether it's better to explore space with men or with machines. Robots have the advantages of being strong, compact, durable, and relatively cheap, but humans have the ability to adapt to any situation and exercise real-time, complex judgments.

The United States thought it had accomplished what it needed to after the excitement and expense of the Moon landings. As NASA funding was dialed back, the agency modified the remaining Saturn V rockets to launch and send astronauts to the Skylab space station. Meanwhile, they began developing a reusable vehicle designed to carry astronauts and equipment into low Earth orbit roughly once a week. The Space Shuttle could carry up to eight astronauts and 25 tons of cargo. Meanwhile, the Soviets gave up on getting men to the Moon after four consecutive failures of their huge N1 rocket; the second exploded on the launchpad with the power of 5,000 tons of TNT. In 1971, they were first to launch a space station dubbed Salyut. But in a chilling example of the hazards of the vacuum of space, the second three-man crew to visit Salyut suffered depressurization of their capsule as they prepared for reentry. They died of asphyxiation in just forty seconds. With the slow thaw in relations between the two superpowers, the Space Race ended. Détente in space was symbolized by the docking of the Apollo and Soyuz spacecraft in 1975 and a historic handshake between Tom Stafford and Alexey Leonov.

The Space Shuttle flew 135 times between 1981 and 2011, sending 300 astronauts into space. In its early years, it was used for a mixture of scientific and military payloads; in its later years, it was used to complete assembly of the International Space Station. It also served as a reminder of the danger and the high cost of space travel.[17]

On January 28, 1986, a nationwide TV audience was stunned when the Space Shuttle Challenger broke up and exploded in a clear blue winter sky, just seventy-three seconds after launch. Later investigation showed that a leak from an O-ring seal on one of the solid-fuel boosters had led to extreme aerodynamic stress on the spacecraft as it traveled at twice the speed of sound. Millions of schoolchildren were watching because NASA had selected Christa McAuliffe to be the first teacher in space. Chillingly, the crew cabin was intact as the vehicle broke up, so the seven crew members most likely died from subsequent impact in the ocean (Figure 13).[18] The grief was repeated seventeen years later when the Space Shuttle Columbia disintegrated as it reentered the Earth's atmosphere at twenty times the speed of sound. During launch, a piece of foam insulation had broken off from the external fuel tank and ruptured the leading edge of the left wing. Fourteen crew members died in these twin disasters.

The Space Shuttle was never as cheap and nimble as planned. Instead of the planned flight each week, the Shuttle managed one flight every two or three months. Over the course of the program, a launch cost about $1 billion, which works out to $80,000 per kilogram placed in orbit. Commercial entities were only able to use the Shuttle with massive subsidies from the government. The US military lost patience with the anemic launch schedule and the fact that two out of five orbiters had been lost catastrophically, so they developed their own heavy-lift capability using rockets without any astronauts.

Figure 13. The loss of the Space Shuttle Challenger seventy-three seconds after launch on January 28, 1986, led to the death of seven crew members. This incident and the death of seven aboard the Space Shuttle Columbia in 2003 are sobering reminders of the hazards of space travel.

However, the Space Shuttle did provide case studies in the importance of having astronauts rather than robots work in space. Robots are not versatile or reliable enough to match a well-trained astronaut. We can all admire the "seat-of-the-pants" problem-solving skills demonstrated by Neil Armstrong as he guided Apollo 11 over a boulder field with hardly any fuel left, and by the Apollo 13 crew as they nursed their crippled spacecraft around the Moon and home to safety. In particular, the five servicing missions of the Hubble Space Telescope defined the state of the art for astronauts, with multiple long space walks, challenging technical jobs, and difficult decisions made under time pressure. NASA administrator Mike Griffin had nixed a final Hubble servicing mission, worried about the risk to the astronauts. But in the end he determined that a robot servicing mission was so difficult that it was destined to fail, so astronauts were called on to give Hubble its final upgrade in 2009.

The choice between robots and humans is a false dichotomy. Machines are pathfinders and advance scouts, learning as much as they can and setting the stage for humans to eventually follow. We've explored the Solar System with robotic probes so far, but they're limited in what they can do. Machines are extensions of us as we explore; when we eventually live in space, they will be our partners.

PART II

PRESENT

Wheel rats. That's what Josefina calls those who never spend time in the Hub. Then we laugh. She's my best friend; I love her mischievous smile and seditious sense of humor. Too many Pilgrims are aloof or self-important; they know we're specially chosen and elite and they often act that way. A few have a messianic streak I find a little scary.

Floating in the Hub, the Earth is a blue-and-white bauble nestled in black velvet. It's cozy and womblike. The Hub is the only zero-g place in the station; all the living and working quarters are around the rim of the wheel, spun to two-thirds g, which avoids the worst problems of bone loss and physiological adjustment. There are no real windows in the rim, since they would reveal the vertigo-inducing view of the Earth wheeling by every thirty seconds. Large panels set into the walls are programmed to display crisp holographic images of forest glades and mountain meadows. To me that's more disorienting, since it's such a disconnect from the reality of being 300 miles high, with only a thin titanium sheath separating us from the frigid, lung-busting vacuum of space.

The dreams still visit me; I can't shake them. By day, I'm consumed with tasks and purpose, but I'm beginning to dread the nights.

The Overseers keep us busy to spare us from dwelling on what we're about to do. The Moon and Mars are home to large colonies, researchers travel routinely as far as Jupiter and Saturn, and robot freighters ply the asteroid belt, but we've never severed the umbilical to the Solar System.

We know the risks. Space is unforgiving and humans are soft and fragile. Amid the high points, there have been disasters. I watched some of them play out as a kid. The orbiting research station was destroyed by a hail of micrometeorites. The first Europa lander was lost due to an orbital miscalculation, flung into deep space. The first Mars colony unraveled due to sectarian rivalries.

I miss my family but can't imagine going back down. Mom and Sis are sharp and clear on-screen, but they've started to sound far off and disembodied. They told us to expect this, the withdrawing. Josefina says she cries most nights and I feel bad for her, then I feel bad I don't feel the same way. The station is a metal carapace and we're shrinking into it to bond with our new tribe.

We're shocked when we hear who'll be kicked off the station. With some we really saw it coming. Rajesh and Dimitri are abrasive and scheming. They've squandered all goodwill with their colleagues. The next to depart are another handful of malcontents, confederates, and henchmen of the ringleaders. There are others about whom we've had our suspicions. They share a haunted look and an inability to make eye contact. They've lost their stomach for the mission, and they have to go because our solidarity and sense of purpose is fragile. But there seems to be no rhyme or reason to the last group. Sonja is among them, and Pierre; we've laughed and shared good times with both of them. However, the profilers have picked them out and there is no arguing with the decision. Some subtle pattern of behavior has marked them as a threat. Josefina and I are on the way to dinner when we see them in the air lock of the shuttle bay. I'll never forget the looks on their faces: angry, sullen, dazed, terrified.

They try to keep things upbeat. The piped music in common areas is soothing or jaunty. They lay on parties and celebrations to vary the routine. Messages from the Overseers are very carefully crafted and positive. And down below? From our vantage point, it's a pretty planet. But the inmates on Earth are in charge of the asylum. All the tools exist to solve the world's problems, but the fractious top species is squabbling and dithering.

Being on the station is in one sense timeless. No change in climate or vegetation gives a hint of the passage of days and weeks. Birthdays and festival days are forgotten or ignored. On the other hand, there's a clear sense of time rushing forward to a vanishing point. That point has nearly arrived.

One evening, Josefina and I go to the Hub and pivot away from the Earth view to the opposite port and the blackness of space. As I float, I reach out and touch her fingertips with mine. Neither of us speaks. Above our heads are three sleek and black obelisks. They float alongside the station, perfectly parallel, poised for our destiny.

Ark 1. Ark 2. Ark 3.

4

Revolution Is Coming

Space Doldrums

NASA has been in the doldrums.

The doldrums are a place, not a state of mind. In the eighteenth century, sailors knew the doldrums as a region near the equator where the prevailing winds might die for days or weeks, leaving sailing ships stranded on a glassy sea. NASA has also been becalmed, and its personnel and its supporters have experienced the accompanying feelings of listlessness and stagnation.[1]

As our story moves from the past to the present, we first describe how far our aspirations have fallen in forty years—from the Moon landings between 1969 and 1972 to an inability to get an astronaut into low Earth orbit. We look at the difficulty of space travel, rooted in the implacable truth of the rocket equation. Then we see a glimmer of hope in the nascent space tourism industry. Last, we draw a parallel between the evolution of information technology and space technology, leading to optimism that resurgence is around the corner.

NASA's lowest point was arguably the 2013 shutdown of the US Government, when 97 percent of its employees were furloughed, the highest percentage of the twenty-four federal agencies. Only a skeleton staff remained to ensure the safety of the crew on the International Space Station. Other activities halted immediately—no research was performed, no missions were planned, no e-mails were answered. It was a stark reminder that the exalted goal of space travel could easily be grounded by terrestrial politics.

The agency has also been struggling with decrepit infrastructure. In 2013, the Office of the Inspector General found that 80 percent of NASA's facilities were more than forty years old and woefully out of date, and carrying maintenance costs of $25 million a year. What's needed is far more than a coat of paint; the backlog of deferred maintenance totals $2.2 billion.[2] NASA's government funding has been shrinking for decades (Figure 14). For perspective, the bank bailout in 2008 cost more than has been spent on NASA since it was started in 1959. No bucks, no Buck Rogers.

Nothing epitomizes the malaise better than the Space Shuttle. By the time of the last flight in 2011, it represented forty-year-old technol-

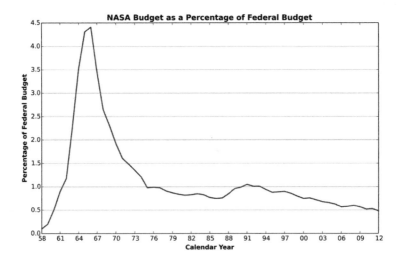

Figure 14. NASA's share of the federal budget since the early 1960s. The rapid buildup for the Apollo program was unprecedented and unsustainable. Since then, there has been a steady decline, apart from a slight rise at the peak of Space Shuttle and International Space Station activity.

ogy. The launch rate ended up ten times lower than originally planned and the cost per launch twenty times higher. Two of the five orbiters suffered a catastrophic fate, with the loss of all on board. Apart from emblematic flights to launch and service the Hubble Space Telescope, most of the time the Shuttle served as an expensive limo to launch satellites and ferry construction materials to another high-priced and outmoded facility: the International Space Station. The Challenger and Columbia disasters are etched in the national psyche, and they have contributed to a widespread ambivalence about America's space program.[3] Since 2011, the United States has been unable to get astronauts into orbit without help from the Russians.

In addition to the fairly frosty relations between the two countries, the Russians have their own problems.

After the fall of the Soviet Union, the Russian space program suffered from diminished budgets and a lack of innovation.[4] In 1965, the Soviets invented the Proton rocket to launch ICBMs, and they still use variations of the original design. In recent years, the Russians have suffered seven mission failures. In 2010, three satellites crashed into the Pacific Ocean. In 2011, a resupply mission to the International Space Station exploded over Siberia in a spectacular fireball, forcing the six waiting astronauts to dig deep into their reserves of food and water. In 2013, another three satellites were lost in an explosion that rained hundreds of tons of toxic debris on the launch site. Yuri Karash, a member of the Russian Space Academy, compared Russian rocket development to attempting to upgrade a steam engine: "You equip it with a computer. . . . You equip it with air conditioning. You put a locomotive driver with a university degree in the cabin, and it will still be the same steam locomotive."[5] The Russian Government audit agency noted that money intended for the space program had simply been stolen.

At the Baikonur Cosmodrome, on the steppes of western Kazakhstan, the decay is obvious. Baikonur is where Sputnik was launched, and where Yuri Gagarin and Laika made history. But today, nomadic herders occupy the many vacant buildings and the town struggles with heroin smuggling and radical jihadists. American, European, and Japanese astronauts arrive at the launch site via a rutted road where camels have the right of way. But they keep on arriving because it's their only way up.

Meanwhile, NASA's generally successful program to send out robotic probes to explore the Solar System is also under stress. The budget for planetary science is falling. Complex interplanetary probes cost a couple of billion dollars each and the budget only has enough slack to fund a couple of missions per decade.[6] A more fundamental problem involves plutonium. Since the 1970s, almost everything we've learned about the outer planets and their moons has relied on power from heat released by the radioactive isotope plutonium-238. Solar power is too feeble and chemical batteries are too inefficient, so this by-product of nuclear reactors (which cannot be used to make a bomb) is the go-to super-fuel. But poor planning and false promises from Russia have left NASA with barely enough plutonium to power missions for the next few years. The nuclear crisis is so bad that affected researchers call it "The Problem."

Communication is another mundane but basic problem. When you watch a silly cat video on YouTube, you give little thought to how the data got to your computer, apart from being dimly aware that the video is really a stream of ones and zeros. In fact, videos and e-mails and data aren't transmitted whole. They're disassembled into packets of data, distributed worldwide via optical fiber and radio waves over a network of networks, and reassembled at your computer or handheld device— rather like digital sausages. It works well for Earth-bound humans, so why would it be hard for an astronaut to watch cat videos on the Moon or Mars?

First, it takes light or radio waves anywhere from four to twenty-one minutes to reach Mars from Earth, depending on where the two planets are in their orbits. NASA engineers don't control the Mars rovers like a video game enthusiast would, flicking a joystick as the rover careens across sand dunes. The rovers are controlled painstakingly by commands that are separated by a half hour or more to allow for the round-trip signal time. Second, planets rotate and shadow the orbiters, so there are dead times when no communication is possible. Third, these interruptions and delays cause technical problems because the Internet paths are in constant flux; if a packet of data sits around too long before its partners arrive, it's discarded. At the moment, the Internet can't be extended into the Solar System. Luckily, the "Father of the Internet" is on the job. Vinton Cerf, designer of the original protocols for the Internet in 1973, is working with NASA on the next-generation system that will operate seamlessly across billions of miles.[7]

However, when one is becalmed in the doldrums, the real problem isn't money or communication. A more fundamental problem is propulsion.

Principles of Flight

Why is space travel so difficult? It's just a matter of accelerating an object to 17,650 mph, as Newton conjectured. But that's as reductive as saying the *Mona Lisa* is just a picture of a smiling woman. A general description gives no sense of the complexity and subtlety of the work.

On the dead calm sea, a breeze springs up, ruffling the water and soothing your fevered brow. As humans have known for millennia, if you can catch this breeze in cloth or canvas, it will propel you forward. Large ships from the Romans through the Vikings used square sails to catch the wind, augmented by men pulling oars. But sailors plying the Mediterranean more than a thousand years ago discovered through experimentation that triangular sails allowed a boat to sail almost into the wind, and this capability was enhanced with multiple sails. Whereas a square-rigger can't move downwind faster than the speed of the wind that's pushing it, a modern yacht can sail several times the wind speed, even when it's almost pointing into the wind.

The explanation was provided in 1738 by the Swiss scientist Daniel Bernoulli, a member of an illustrious family of mathematicians and scientists. The physical principle states that in any fluid flow, an increase in the speed of the fluid is accompanied by a decrease in pressure. Wind forced to travel over the curved surface of a sail must travel faster than wind moving behind the sail; the decrease in pressure on the front face of the sail creates a force that drives the boat forward.

Now imagine that the sail is horizontal. If it can be propelled through the air, it will experience that same force in an upward direction. The principles of flight are based on Newtonian physics, refined over several hundred years.

A flying object such as a bird, a plane, or a rocket is engaged in a constant tug of war among opposing forces. The downward force is the inescapable foe: gravity. The upward force is lift, provided by air flowing over a wing. The forward force is thrust, provided by muscles for birds and engines for planes. It's opposed by drag, the resistance from the air, which can be minimized by careful aerodynamic design.

Human flight began with balloons. With a balloon, thrust comes from the whims of the wind and lift comes from the buoyancy of a gas less dense than air. The Chinese developed hot-air balloons for military signaling in the third century, at the same time that they were developing "fire arrows." In 1783, Jean-François de Rozier and the Marquis d'Arlandes became the first humans to fly, traveling five miles across the French countryside in a balloon designed by the Montgolfier brothers. They had to petition King Louis XVI for the honor, since he had originally decreed that condemned criminals would be the first test pilots. Balloons reach their limit at the height where even the lightest gas, helium, can't provide buoyancy in the thin air. Austrian daredevil Felix Baumgartner got close to this limit in 2012 when he ascended to 24 miles in a balloon that was three times the height of a commercial jet. He took the quick route down, leaping from the balloon in a pressurized suit. In his four-minute-long free fall, he broke the sound barrier and reached a speed of 844 mph.[8]

Powered flight began modestly in 1903 when Orville Wright traveled 120 feet just a few feet off the ground at slower than running speed. The Wright brothers observed birds and conducted many experiments on wing shape and profile. A flat wing can provide lift, but modern airfoil design has led to a curved upper surface, echoing birds and boats. Their plane was built using spruce wood, a bicycle chain to drive the twin handmade propellers, and a custom-built engine, since no existing automobile engine was suitable. The brothers tossed a coin to decide who would make the historic first flight.

Throughout the twentieth century, airplanes traveled faster and higher. Thrust came first from variations on the automobile's internal combustion engine, which was used to drive a propeller. Aircraft like this reached altitudes of 10 miles and speeds of 450 mph by midcentury, but they began to be supplanted by jets. The jet engine was the brainchild of RAF Officer Frank Whittle, who overcame significant physical limitations to become a pilot. His innovation was an engine that took in air, compressed it in a turbine, combusted the air–fuel mixture, and ejected the burning gas at high speed through a nozzle. This type of engine is most efficient at high speed and high altitude. Jets pushed altitude and speed records to the dizzying heights of 35 miles and 2,190 mph, or more than three times the speed of sound.[9]

The quest for space brings us back to the uneasy relationship between civilian efforts motivated by exploration and the shadowy world of the military. For example, the top speeds of military aircraft such as the SR-71 "Blackbird" are classified. The US Air Force has built a series of aircraft whose existence was not acknowledged by the government, military personnel, or defense contractors. Examples of these "black projects" include the Mach 3 Blackbird, the F-117 Nighthawk

Figure 15. Schematic view of the layers of the Earth's atmosphere. Space is typically demarcated by the Kármán line at 100 km, where the atmosphere is too thin to support aerodynamic flight. Low Earth orbit is any altitude ranging from 160 to 2,000 km.

stealth aircraft, and the B-2 bomber. All of these are air-breathing jet aircraft, incapable of reaching space.

Jet engines can't work beyond 100 kilometers or 62 miles, where air is two million times thinner than air at sea level. This boundary is called the Kármán line. At that altitude, an airplane would have to move at orbital velocity to generate enough lift to stay aloft (Figure 15). Space has no edge. The atmosphere thins out gradually into a perfect vacuum. Low Earth orbits start around 100 miles up, below which the tenuous atmosphere would create enough drag on a satellite to make it descend and burn up. The International Space Station orbits at the magisterial altitude of 250 miles.

However, the military did have a series of projects—some of which were secret—involving planes powered by rockets. The experimental X-planes began with the Bell X-1, which had its maiden flight in 1946. On October 14, 1947, Captain Chuck Yeager became the first human to travel faster than sound, flying in an X-1 over the Mojave Desert in California. When the embargoed news appeared in *Aviation Week* and the *Los Angeles Times*, the journalists involved were threatened with prosecution, although that didn't occur.[10] Yeager never went to college but he rose to the rank of brigadier general, piloted dozens of experimental aircraft, and set a speed record of Mach 2.4. Soon after reaching that unprecedented speed, his X-1 became violently unstable and tumbled 50,000 feet in fifty seconds. Yeager regained control just in time for a normal landing. His laconic, understated style was branded in the popular culture as "the Right Stuff."[11]

In 1959, the Air Force rolled out the X-15, the fastest plane ever built. It was a spaceplane, a vehicle that worked like an aircraft in the Earth's atmosphere and like a spacecraft when it was in space. The descent was an unpowered glide to landing. In 1967, Lieutenant Pete Knight took the X-15 to 4,520 mph (Mach 6.7), a record that still stands nearly half a century later. Like Chuck Yeager, he had his share of close calls; on one flight, he lost power to all onboard systems and had to bring the X-15 down from an altitude of 173,000 feet for an emergency landing.[12]

The Air Force and NASA collaborated on developing the X-15, but they diverged when the young space agency chose the Atlas and Redstone rockets for the Mercury program that put the first Americans in space. Eight Air Force pilots flew the X-15 high enough to earn their astronaut wings, including Neil Armstrong, who would later become the first man to set foot on the Moon.[13] Only three X-15s were built.

They flew 199 missions. One pilot was lost when his plane went into a hypersonic spin and broke apart at 60,000 feet, scattering wreckage over 50 square miles.

The only other spaceplanes so far have been NASA's Space Shuttle, its Russian counterpart Buran (which flew once, in 1988), Burt Rutan's SpaceShipOne (which flew seventeen times between 2003 and 2004 and is discussed in chapter 5), and the X-37. The X-37 is a project to demonstrate reusable space technologies. It began in 1999 under the Air Force and was transferred to NASA in 2004. Think of it as a more advanced but smaller and unmanned version of the Space Shuttle. Like the Shuttle and unlike the other American spaceplanes, the X-37 is a true orbital vehicle that can reach an altitude of at least 100 miles.[14] It has flown just three times. In 2011, Boeing announced plans for a scaled-up version that would carry up to six astronauts in a pressurized compartment in its cargo bay.

Since the X-37 is a black project, there's only speculation about what it does in orbit. In an amusing geopolitical irony, the Atlas rocket that launches the X-37 uses the sturdy Russian RD-180 engine for its first stage. Early in 2014, amid tensions over the situation in Ukraine, the Russian defense minister announced that Russia would no longer supply rocket engines for US military launches. Boeing will switch from the Atlas to the American-made Delta family of rockets, which also have more than a million pounds of thrust.

It sounds simple: Get a big and powerful enough rocket and you can escape the doldrums. But there's a catch, which brings us back to Konstantin Tsiolkovsky.

His equation says that the final velocity of a rocket depends only on the exhaust speed of the fuel and the ratio of the fuel mass to the payload mass. No matter how cleverly you design your rocket, no matter how ingenious your engine, you are governed by the rocket equation. The bad news is that the fuel-to-payload ratio is tucked inside a logarithm. If you increased the amount of fuel by a factor of a thousand, it would buy you only an extra factor of seven in the rocket speed. The more fuel you pack onto a rocket, the more energy the fuel needs to waste pushing the rest of the fuel. The highest exhaust speeds of chemical fuels are around 4,000 meters per second or 9,000 mph. So the rocket equation says that to reach orbital speed requires ten times as much fuel as payload. Many dreams have been dashed on the rocks of the rocket equation.

Space Tourism

By never sending a poet or artist into space, NASA missed a big opportunity to engage the public with the excitement of space travel.

Any creative spirit would be inspired by the experience of being weightless, or of looking in one direction at the gossamer-thin edge of the Earth's atmosphere and in the opposite direction at the inky blackness of endless night. Poets have the ability to describe the indescribable, so they would have given those of us with feet rooted to the ground a sense of why we might want to leave Earth's cradle.[15]

At the dawn of the Space Age, NASA considered sending acrobats and contortionists into space, as well as women, since they were smaller and lighter than men. But the United States was in the Cold War and President Eisenhower specified that astronauts must be military test pilots. This decision simplified the selection procedure because all of the five hundred men who applied were highly disciplined alpha males. Each member of the first group selected in 1959 was under forty, stood less than five feet eleven inches tall, had degrees in engineering, and had logged at least 1,500 hours of flying time as jet test pilots. The degree requirement excluded accomplished Air Force test pilots like Chuck Yeager, who had enlisted young and come up through the ranks.

The result was good-natured but intense rivalry. Mercury astronauts were national heroes, while the test pilots pointed out that the astronauts didn't actually fly the spacecraft so they were just "Spam in a can."

The result was a monoculture of astronauts who were male, white, unflappable, and typically from the Midwest.[16] Referring to the twenty-four astronauts from Ohio, TV personality Stephen Colbert asked Ohio congresswoman Stephanie Tubbs Jones, "What is it about your state that makes people want to flee the Earth?"[17]

The selection vise was loosened slightly for the second intake in 1962, when civilian pilots were included, letting in Neil Armstrong among others. In 1965, degrees in medicine and science were accepted and the flight requirement could be met after joining the corps. But by far the biggest change came with the eighth intake in 1978, when NASA divided astronauts into two categories: pilots and mission specialists. That group of thirty-five included six African Americans, one Asian American, and six women, one of whom was Sally Ride, America's first female in space.

Meanwhile, the demographics of Soviet and Russian cosmonauts were also narrow, except for the textile factory worker and recreational skydiver Valentina Tereshkova. In 1963, Tereshkova became the first woman and the first civilian to fly in space.

When NASA announced its Teacher in Space Program in 1984, Christa McAuliffe was selected from the 11,000 teachers who applied. There was also a Journalist in Space Program—applicants included Walter Cronkite and Tom Brokaw. An Artist in Space Program was even being considered—but then came the stunning Challenger Space Shuttle disaster, which killed McAuliffe and her six colleagues. NASA quietly ended its civilian space program, although McAuliffe's backup, Barbara Morgan, did eventually fly on the Space Shuttle, after she had retired from teaching and joined NASA's astronaut corps.

As of late 2013, 540 people from thirty-eight countries had reached low Earth orbit or beyond. They ranged in age from twenty-five to seventy-seven. Being in space remains very special; for comparison, more than 4,000 have stood on the summit of Mount Everest. The vast majority of space travelers are employed by the military or government agencies, but the past decade has seen the emergence of a new industry: space tourism.[18]

After the loss of a second Space Shuttle in 2003, all remaining flights went toward NASA's fulfillment of its commitments to finishing the International Space Station. Missions involving scientists or civilians were shelved. Meanwhile, the fall of the Soviet Union left the Russian space program debilitated and starved for cash. So the Russians eagerly agreed when the Tokyo Broadcasting System offered to pay them $28 million to send one of its reporters to the Mir Space Station in 1990. In 1999, MirCorp was formed to use the aging Russian space station for tourism. It was funded mostly by American entrepreneurs. MirCorp partnered with a Russian launch company to boost Mir into a higher orbit and it signed an agreement with NBC and Mark Burnett, who had recently produced the *Survivor* TV series. American engineer and millionaire Dennis Tito was announced as the first self-funded space tourist. NBC even ran ads for its upcoming *Destination Mir* reality TV show.

But trouble was brewing. NASA officials and members of Congress heavily criticized MirCorp for interfering with international space treaties and for trivializing space exploration. It was also awkward for NASA that the cut-price Mir program was exposing the enormous cost of NASA launches and the International Space Station. NASA pressed the Russians into de-orbiting Mir and tried hard to squash the nascent space tourism industry, but Tito ignored the furor and traveled by Soyuz spacecraft anyway, spending eight days on the International Space Station in 2001. He was followed by a South African, an American, and an Iranian American woman. In 2009, Hungarian-born businessman Charles Simonyi became the first repeat space tourist. By the time Cirque du Soleil founder Guy Laliberté became a space tourist later in 2009, the price had gone from $20 million to an eye-popping $40 million.[19]

We can admire astronauts, but it's hard to empathize with them. They are so confident and accomplished that normal insecurities are hidden from view, and their sensibilities are analytic. The first space tourists have all been wealthy entrepreneurs—the 0.01 percent—and they're not like us either.

NASA briefly gestured toward conscripting artist astronauts in 2003, when they selected performance artist and musician Laurie Anderson as the first (and last) Artist in Residence.[20] It didn't go well. On her first day, as she was sitting in the office of the director of the Johnson Space Center, she asked, "When do I go up?" He told her that wasn't going to happen.

A Striking Parallel

Is there any way to predict the future of the space program? Its progression so far, with its fits and starts, iconic visionaries, and incubation during the Cold War, seems unique. But its history has striking parallels with an indispensable component of modern life: the Internet. Both were incubated by the military-industrial complex, both grew thanks to government investment, and both have gained striking new capabilities courtesy of the involvement of entrepreneurs from the private sector, whose investment potentially dwarfs that of the government (Figure 16).

Robert Goddard and Wernher von Braun are to the rocket as Joseph Carl Robnett Licklider is to the Internet. His name is known only to the most passionate computer geeks, but he deserves a nod of appreciation from anyone who has effortlessly viewed a web page or sent an e-mail or an image halfway around the world in under a second. Known as J. C. R. or "Lick" by his friends and colleagues, he was a psychologist who liked to work on refurbishing cars in his spare time. Licklider was a

THE INTERNET

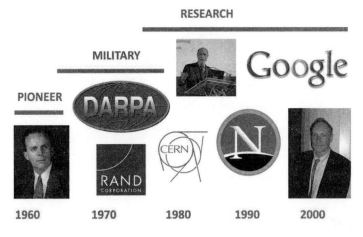

Figure 16. The Internet had pioneers who foresaw a worldwide interconnected communications system. It was then incubated by the military and in research labs before it emerged into the private sector. Research by the private sector now drives the development and innovative use of the Internet.

professor at MIT in the 1950s, working on the physics of sound perception, when he became interested in computers. At the time, computers were rare, expensive behemoths the size of a bus that used as much electricity as a small town yet were far less powerful than your smartphone. The only way to transmit digital data was to write it onto bulky magnetic tape and send it in the mail.

Yet Licklider had an uncanny sense of the potential of the information age. He foresaw such capabilities as graphic displays and point-and-click interfaces, e-commerce and online banking, and digital libraries. The only mass media at the time were TV and radio, yet he visualized the two-way flow of data and information over a worldwide network of computers. And he predicted that software would be downloaded as needed from that network. Licklider wasn't an inventor, but he was a prodigious source of ideas. He formed and funded research groups that began to develop the capabilities we now take for granted. Lick could project the worldwide potential of his primitive information technology just as Goddard could project the eventual capabilities of rockets, even though his first rocket flew less than 200 feet.

Like the space program, the Internet depended on investment by the military to mature. The US military establishment was particularly concerned about moving data efficiently between command centers and having redundancy and resilience in the case of a nuclear attack. In 1962, Licklider was hired by the Department of Defense to work at DARPA, the Defense Advanced Research Projects Agency. On October 29, 1969, a real-time link was established between research labs at UCLA and Stanford. Three characters were sent before the system crashed, but this simple transmission, reminiscent of the first phone message by Alexander Graham Bell almost a century earlier, was the start of a revolution.

DARPANET was the technical core of what would become the Internet. By the early 1980s, a new node was being added every twenty days. Many technical problems were solved in these pioneering years, such as designing protocols for chopping data into packets, sending them on diverse paths through the network, and seamlessly stitching them back together at the destination. The second phase of Internet development was carried out by universities and government labs. By the end of the 1980s, the National Science Foundation (NSF) laid down the physical backbone of a high-speed Internet and NASA was providing connectivity to more than 20,000 scientists across seven continents.[21]

At that time, the public was unaware of the Internet. It was used by researchers to send data and e-mail. Commerce was forbidden.

The floodgates opened in the mid-1990s. Private Internet Service Providers (ISPs) had sprung up to meet growing public demand for e-mail access. The US Congress passed a law that allowed the NSF to support access to networks that weren't used exclusively for research and education. This created angst as researchers worried that the new Internet might not be responsive to their needs. The online world had always been a geeky place of text and equations, but in 1989 CERN

researcher Tim Berners-Lee released his hypertext concept for public
use. In 1993, a team led by Marc Andreessen at the University of Illi-
nois increased the visual appeal of the Internet by releasing the first
web browser, called Mosaic. Encryption was added soon afterward to
make transactions more secure.

In 1995, the NSF dropped all restrictions on Internet commerce
and let private companies take over the high-speed "backbone." That
year also saw the founding of the Yahoo search engine, the auction site
eBay, and the online bookseller Amazon. Huge new audiences adopted
the technology and used it in unforeseen ways. As we will soon see, the
space industry may now be where the Internet was in 1995, ready
to soar.

The core of the analogy is that the government and the military
have deep enough pockets to develop technology with no eye on profit
or return on investment. Once the field has been prepared and tilled,
the private sector can scatter seed and see what grows best.

With too much government and military control, technologies can't
reach their full potential. President Dwight Eisenhower used his fare-
well address to warn of the dangers of the "military-industrial complex."[22]
It's ironic that this five-star general and two-term president—the quintessential Washington insider—issued such a clarion call against concentration of influence within and around the government. He said: "We
must guard against the acquisition of unwarranted influence, whether
sought or unsought, by the military-industrial complex. The potential
for disastrous use of misplaced power exists, and will persist."[23] The
analogy between access to space and access to information seems to
break down. However, the connection is uncanny when we recall the
current controversy over the highly sophisticated and intrusive harvesting of personal data over the Internet by the US Government.

To understand the potential of space tourism, it's helpful to look at
the growth of the Internet. Since the Internet entered commerce and
culture, its rise has been meteoric. It accounted for one percent of two-

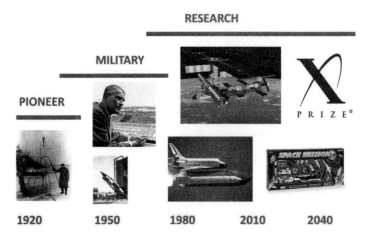

Figure 17. The space program also had visionaries who aimed for a permanent human presence in space. Progress was spurred by a military superpower rivalry and fostered by NASA. Private investment has recently begun so the space industry sits now where the Internet was in the early 1990s.

way telecommunication traffic in 1993, but that rose to 50 percent in 2000 and 99 percent today. In 1993, there were a million Internet hosts; now there are a billion. Space travel is poised to follow the trajectory of the Internet, becoming demilitarized and then massively commercialized (Figure 17). Leaving Earth may soon be cheap and safe enough that it becomes an activity for the masses rather than the experience of a privileged few. Some of the recently formed space companies will be like Netscape and Altavista—the web-browser and search-engine leaders in 1995 and now long forgotten—and some will become behemoths like Google. The next decade promises to be very interesting.

5

Meet the Entrepreneurs

The Radical Designer

Entrepreneurs are like the high-octane fuels needed to take space travel to the next level—volatile and sometimes hard to handle, but capable of unprecedented performance. Technical pioneers work best when they're unfettered by conventional wisdom or institutional constraints. They have their eyes set on ambitious goals that might sound quixotic, but they pursue those goals with breathtaking passion and relentlessness. If they're outsiders with modest means, they need deep pockets behind them to achieve their goals.

We've seen this combination of ingredients with Robert Goddard. He did his pioneering experiments while being shunned by academia and scorned by the military. His early work was sponsored by a modest grant from the Smithsonian Institution. Then came Harry Guggenheim, son of Daniel Guggenheim, who owned mining companies and, by the end of the nineteenth century, had one of the largest fortunes in the world. Harry, a former Navy pilot and president of the family foundation, was a friend of Charles Lindbergh, who introduced him to Goddard. In 1930, Lindbergh received a Guggenheim Foundation grant of

$100,000, which would be worth $4 million today.[1] This support helped launch the Rocket Age.

Goddard was so far ahead of his time that his work was unregulated. The Federal Aviation Administration (FAA) was formed in 1926, and it wasn't until 1984 that the agency had a division to oversee rockets and commercial space travel.

Burt Rutan has an entrepreneur's impatience with red tape.

When asked how he approached the Federal Aviation Administration about launching into space from a remote new site in the Mojave Desert, Rutan said, "It's better to ask for forgiveness than permission." A lifelong pilot, he's now in his early seventies and has heart problems. He calls the defibrillator implanted in his chest a "standby ignition system." Alluding to his health issues, he said he's discovered that when you're in an airplane and you push the throttle forward and pull the stick back, it will take off even without a medical certificate.[2] So far, he's never been grounded for these infractions of regulations.

Rutan has worked his way into space from the ground up. He grew up in rural Oregon in a house with no plumbing, and his parents followed a religious sect that prohibited activity on the weekend. Unable to play sports, he started making his own model planes at the age of eight and developed an intuitive feel for design. "I never built from a kit," he recalled. "I bought balsa wood and invented a new airplane." He felt that keeping his creative side in the foreground made him a better engineer later in his career.

Rutan was hired straight out of college in 1965 as a civilian flight test engineer for the US Air Force. His job was to solve stability problems with the F4 Phantom jet fighter, which had suffered sixty-one crashes. The work became personal when he witnessed the death of his friend Mike Adams in an X-15 crash, also because of stability problems. Rutan invented a spin recovery system that prevented the F4 fleet from being grounded. Many of his homebuilt designs would use *canards*—small wings located ahead of and slightly above the main wings to give greater control and stability. Rutan also liked to employ a second, "pusher" engine at the back of the airplane, and he was an early adopter of light, composite construction materials.

At thirty, he started his first company, the Rutan Aircraft Factory. The two-seaters he designed were used by everyone from hobbyists to NASA. Instead of metal, his kits used foam and fiberglass. When asked how long it took to build an airplane, his pithy response was "one and a half wives." His recreational planes are masterpieces of efficiency and sophistication, but Rutan was looking for a bigger challenge. In 1982, he founded a new company, Scaled Composites, and for thirty years the cutting edge of aircraft design was located "under his wing" in the arid, lunar landscape of Mojave, California.[3]

Rutan caught the world's attention with Voyager, the first airplane to circumnavigate the world without refueling, which was considered impossible by many aeronautics experts. Flying around the world without refueling is like getting to orbit in one key aspect: most of the weight is fuel. The Saturn V rocket was 90 percent fuel as it launched and Voyager was 73 percent fuel when it took off. There was barely room for a pilot and copilot, and the ability of the crew to endure the flight was considered the biggest risk of failure.

Voyager looked like a dragonfly, and it was effectively a flying gas tank, with fuel filling the wings and the spars. Rutan used a radical design for the airframe and wings, where paper honeycomb was sandwiched by graphite fiber composite. Weight was reduced ruthlessly. The most important statistic in aerodynamics is the lift-to-drag ratio: the higher the better. Voyager had a lift-to-drag ratio of 27, better than a jumbo jet (17) or an albatross (20). He came up with the concept while having lunch with his brother Dick in a Mojave diner, sketching it on a napkin. Rutan told his staff to throw every new part up in the air for a weight test, and "if it comes down, it's too heavy."

Rutan had no capital behind him, so he built the plane on the cheap. Company after company turned him down for sponsorship. The owner of Caesars Palace in Las Vegas was willing to fund him, but only if the plane took off and landed from the casino parking lot, which was far too small for the job. One firm wanted to charge him $50,000 to fabricate the wings, so the team figured how to do it themselves for a few hundred dollars. Instead of wind-tunnel testing, Rutan flew a model on top of his Dodge Dart station wagon. He said that wind tunnels only tell you what you already know. He used his own money to finish the project. In December 1986, Dick Rutan and Jeana Yeager, daughter of famed test pilot Chuck Yeager, took off on their perilous journey. They landed nine days later with just 100 of their 7,000 pounds of fuel left.[4]

For his next challenge, Rutan was slightly better funded.

He was drawn to the challenge of suborbital space flight because, as he put it in a 2010 interview, "We can achieve some breakthroughs by making such flight orders of magnitude safer and orders of magnitude more affordable."[5] He has noted that in 1961 Alan Shepard flew into space in a small capsule and ten years later was golfing on the Moon. Progress in that decade seemed unstoppable. He thinks that if you'd told someone in 1971 that now we'd be buying rides into space from the Russians, it would have seemed like heresy.

In the late 1990s, Rutan approached the billionaire Microsoft cofounder Paul Allen with the idea of competing for the Ansari X Prize. A California foundation had offered $10 million to the first organization to fly a manned spacecraft 100 kilometers high twice in a two-week window. Rutan wanted to avoid the complications of a rocket launch from the ground by using a large airplane to carry the rocket to a moderately high altitude and then letting the rocket do the rest. Landing, however, was a challenge. He wanted to avoid an unguided parachute descent and he preferred not to use the heavy heat shields employed by the Space Shuttle and Soyuz vehicles.

His clever solution was inspired by the way a badminton shuttlecock automatically orients itself correctly with the direction of flight. Allen

First Spaceplanes

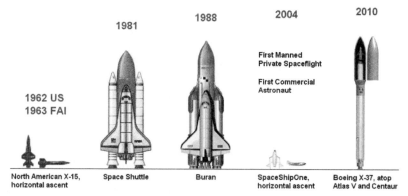

Figure 18. Spaceplanes over the past half century. The X-15 was an experimental jet of the US Air Force; then the 1980s saw the US and Russian versions of a rocket-borne shuttle. Burt Rutan's SpaceShipOne was a landmark, the first successful private venture into near space, and the Boeing X-37 is a new rocket-borne spaceplane.

and Rutan became partners, and SpaceShipOne started taking shape in the California desert (Figure 18). In keeping with his ethos of intuitive, hands-on engineering, Rutan tested the stability of SpaceShipOne by throwing a model off a tower. In June 2004, a crowd of 10,000 people watched Rutan's mother ship, White Knight, haul SpaceShipOne up into the sky. It became the first manned civilian vehicle to reach an altitude of 100 kilometers. In September of that year, SpaceShipOne won the X Prize with two flights five days apart. The only sour note came with an argument between Rutan and Allen—the investor wanted the press but not the public to see the launch. Rutan wanted to inspire the next generation to do great things. He prevailed, and sixty school buses loaded full of kids saw the historic flight. Perhaps this provided the spark for the next generation of space entrepreneurs.[6]

Unassuming and soft-spoken, Rutan is one of the foremost space innovators of our time. He's created nearly 400 aircraft designs. His planes have broken dozens of records and five are displayed at the Smithsonian National Air and Space Museum in Washington, DC. In

2008, he took another step in his progression as a space pioneer when he formed an alliance with a British billionaire who is as expansive as Rutan is self-contained.

The Media Mogul

Richard Branson owes a lot to his mum. To her dismay, he was dyslexic and withdrawn as a child, refusing to talk to adults and clinging to her skirt. To break him of these habits, she once stopped the car three miles from home and let him out. At age seven, he had to talk to strangers to find his way home. He made it, though it took him ten hours. This harsh treatment made him more comfortable talking to adults.

Then she stepped in again when he was twenty-one, to save him from significant jail time. Branson started a magazine called *The Student* just before he dropped out of school and then began a mail-order record business he called Virgin, running both operations from the crypt of a church. He opened his first Virgin record store on Oxford Street in London in 1976 but had major cash-flow problems. To pay off a bank loan, he pretended to buy records for export to evade an excise sales tax. He was arrested, spent a night in jail, and was able to avoid a trial after his mother remortgaged the family home to pay the settlement. He emerged from the episode chastened and emboldened to do better. As he noted in his autobiography, "It is unlikely, not to say impossible, that someone with a criminal record would have been allowed to set up an airline."[7]

Meet Sir Richard Charles Nicholas Branson, disarming yet self-serving, humble yet acquisitive, charming yet brash, founder of an empire of 400 companies, and the fourth wealthiest person in Britain. He has all the hallmarks of ADHD and it's a good bet he has the explorer gene.

Branson cut his teeth selling records, but he operates like a butter-fly collector, or a butterfly, flitting from flower to flower. He's dabbled in everything from condoms (his Mates brand failed) and mail-order brides (he couldn't get any customers) to booze (Virgin Vodka) and pulp fiction (Virgin Comics). He's evolved a clever form of branded venture capitalism where the Virgin Group acts as a loose umbrella and the brands are all leveraged by his gift for marketing, yet each one is free to experiment and fail. He's even branded his intuitive, freewheeling management style, expounded on it at length in a 600-page autobiography titled *Losing My Virginity*. Branson laughs when he says, "I don't complicate my life with financial reports," and he claims not to know the difference between net and gross profit.[8]

Branson has eclectic interests, but he recognized the aviation and space industries as particularly ripe for innovation. He left the safe harbor of a highly profitable music selling and recording business to go into the risky commercial airline business. Virgin Atlantic started in 1984 with a single jumbo jet leased for a year. He nearly failed at the outset when birds flew into one of the engines during a certification flight and he didn't have the million dollars needed to replace it. He managed to borrow the money and a few days later had an inaugural flight that became a transatlantic party with free-flowing booze and topless models. He stuffed the flight with journalists to guarantee good publicity and to burnish his reputation as capitalism's fun-loving wild child. Then he had to cope with a long and brutal fight with government-subsidized British Airways. His adversary used such dirty tricks as impersonating his staff, hacking his passenger lists, and spreading lies about him and his company.[9] Branson sued for libel and won a billion dollars in an out-of-court settlement, but rising fuel costs and an economic downturn made running an airline in the early 1990s difficult. In a decision that he said broke his heart, he sold his music business to keep the airline afloat. Characteristically, he reacted to his troubles by reaching even higher, getting into a challenging business that had no track record at all—space travel. He said he was inspired to think about space travel by a question he was asked on a BBC children's TV show in 1988.

Branson founded Virgin Galactic in 2004 and then commissioned Burt Rutan to scale up his SpaceShipOne design to be suitable for space tourism. Whereas SpaceShipOne had one pilot, SpaceShipTwo carries two pilots and six passengers. The carrier aircraft, White Knight II, will take off from a custom-built facility in New Mexico, with a 10,000-foot runway and a suitably "space age" terminal building. At an altitude of 52,000 feet, SpaceShipTwo will rocket upward to just over 100 kilometers, or 60 miles, where the curved limb of the Earth will be visible and the sky will be jet black. The total flight time will be two and a half hours, with just six minutes of parabolic weightlessness at the top of the arc of its trajectory. For this experience, Virgin Galactic is asking a cool quarter of a million dollars (Figure 19).

They're getting it. As of late 2013, more than 650 people had paid deposits totaling $80 million. The list of people on the reservation list included Brad Pitt, Tom Hanks, Katy Perry, Paris Hilton, and Stephen Hawking. Branson, a diehard Trekkie, named his new spacecraft the Enterprise. He asked William Shatner to go up but said Shatner

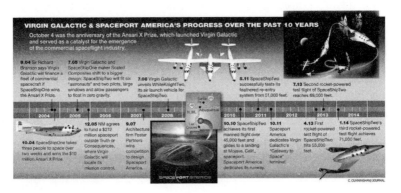

Figure 19. The timeline of Virgin Galactic begins with Burt Rutan's victory in the X Prize competition with SpaceShipOne in 2014, and the selection of a site in southern New Mexico as the launch facility for SpaceShipTwo flights. Progress was put on hold by a fatal accident and the loss of SpaceShipTwo in late 2014.

declined because he was afraid of flying. Shatner's version is that Branson asked him how much he'd pay to go on the inaugural flight and he replied, "How much would you pay *me* to go on it?"[10]

Branson has said that White Knight II represents "the chance for our ever-growing group of future astronauts and other scientists to see the world in a new light." He thinks humanity will have to spread beyond the planet in order to prosper.

However, even for pioneers with the magic touch like Branson and Rutan, the space business is risky. In 2007, three people were killed and another three injured in an explosion at Rutan's Scaled Composites factory, and a year later Rutan said, "Don't believe anyone who tells you the safety will be the same as a modern airliner's." SpaceShipTwo reaches a top speed of 2,500 mph and passengers pull 6 g's on the way down. They wear helmetless spacesuits, which could be a problem if the spacecraft loses pressure at 300,000 feet. In more than thirty test flights, only three have been at supersonic speed. Early in 2014, Virgin Galactic switched to a new, plastic-based, solid rocket fuel, and in October a pilot was killed and another seriously injured when the SpaceShipTwo rocket malfunctioned. This will add to the delay of the first commercial launch, already totaling five years.

Branson's never been stuck behind an executive desk—he's a hands-on adventurer. In 1986, he raced a boat across the Atlantic faster than anyone had before. The next year, he was first to fly a hot-air balloon across the Atlantic. In 1991, he broke both distance and speed records crossing the Pacific, also in a balloon. When SpaceShipTwo finally has its inaugural flight, Branson and his two adult children, Holly and Sam, will be on board.

The Space Futurist

"Over the next 20 to 30 years, humanity will establish itself in space, independent of Earth." Peter Diamandis is sublimely confident that the teething problems of the private space industry will soon be over and we'll be on our way to becoming an interplanetary species. This isn't a goal mentioned anywhere in the Space Act that guides NASA. As he put it, "Not since lungfish crawled out of the oceans onto land has this happened!"[11]

Like Richard Branson, Peter Diamandis is a serial entrepreneur and a Trekkie. He started young. By the age of eight, he was lecturing friends and family on the space program and at twelve he won first place in a rocket design competition. As a sophomore at MIT, he founded a space organization for students that spread nationally. He went to Harvard Medical School—mostly to please his parents, who were both medical professionals—but the pull of space was stronger. Before attaining an MD, he started the International Space University, an institution that has produced 3,300 graduates and has a $30 million campus in Strasbourg, France. He also started a company named International MicroSpace to launch microsatellites, winning a $100 million contract from the US Defense Department. But Diamandis was overstretched and the company couldn't deliver on the contract, so he sold it (for a fat profit, naturally).

In 1994, Diamandis read Charles Lindbergh's memoir *The Spirit of St. Louis* and was inspired. He learned that a hotel owner named Raymond Orteig had put up a $25,000 prize in May 1919 for the first nonstop airplane flight between New York and Paris. Nine teams spent a total of $400,000 trying to win. Lindbergh was considered a dark horse, an outsider with no backing and little aviation experience. He had quit college and spent his early twenties as a "barnstormer," crisscrossing the country in a small biplane giving joy rides and doing aerobatics. He spent a year flying for the Army and another year flying mail for the US Post Office Department. He heard about the Orteig Prize and was immediately interested.

Six famous aviators had died trying to win the prize by the time Lindbergh entered the competition. He had never even flown over a large body of water. The press dubbed the twenty-five-year-old "the flying fool." He made the epic flight from New York to Paris in thirty-three hours, with his fuel-laden plane barely clearing the telephone lines at the end of the runway, having to deal with fog and ice along the way, and landing by dead reckoning on the darkened Le Bourget airfield.[12] Flying was still dangerous when Lindbergh won the prize in 1927, but excitement and heightened interest propelled a new industry. It had taken eight years for the prize to be won, but within three years passenger air traffic increased thirtyfold. The prize spawned the $250 billion aviation industry.

Diamandis had found his business model.

The cost of going into space hadn't changed for thirty years, so it was the natural target for his "incentive prize." When he announced the X Prize, he had no funding and most of the people to whom he pitched the idea thought he was crazy, which only encouraged him. After five years, he persuaded the Ansari family to fund a $10 million prize—competition spurs innovation (Figure 20). The head of the family is Anousheh Ansari, who was born in Iran and trained as an engineer. She moved to the United States after the revolution in 1979 and founded a series of telecommunications companies.

Seven organizations spent $100 million trying to win the prize. Burt Rutan succeeded in 2004, and his SpaceShipOne now hangs above Apollo 11 and next to Lindbergh's Spirit of St. Louis in the Smithsonian. Two years later, Anousheh Ansari herself became the fourth space tourist when she traveled to the International Space Station aboard a Russian Soyuz spacecraft.

Since then, X Prize challenges have proliferated. After giving a talk

Figure 20. The X Prize and emerging commercial space ventures are keeping NASA on its toes. This prototype for a lunar electric rover is designed to support a future lunar base. It will allow two astronauts to eat, sleep, and travel for two weeks, and the rover will be able to cover thousands of miles and navigate slopes of up to 40 degrees.

at Google, Diamandis was approached by a guy in a T-shirt who said, "Let's have lunch." It was Larry Page, the Google CEO, who has since helped extend the scope of the X Prize to address humanity's broad challenges in health, energy, and the environment. In addition to Larry Page, the board of trustees includes film director James Cameron, media guru Arianna Huffington, and astronaut Richard Garriott.

The current competition Diamandis is most excited about is the Qualcomm Tricorder X Prize, awarding $10 million to a working version of Dr. McCoy's medical device from *Star Trek*. By talking to the device, coughing into it, or doing a skin prick, it will measure vital functions and diagnose fifteen diseases more accurately than a board-certified doctor. A true tricorder device hasn't been developed, but there has been a recent explosion of health-monitoring and diagnostic apps for smartphones. "Ultimately this is about democratizing access to health care around the world," says Diamandis, and he notes that, as with space travel, "The technology is evolving much faster than the regulations are."[13]

The only "failed" competition was the Archon Genomics X Prize to accurately sequence 100 genomes in ten days or fewer, at a cost of less than $1,000 per genome. In that case, burgeoning progress in the biotech sector rendered an incentive prize moot.[14]

The quintessential experience of an astronaut is zero gravity. To whet people's appetite for space, Diamandis founded a for-profit company called the Zero G Corporation to give paying customers a taste of weightlessness in parabolic flight. It was 1992, the 500th anniversary of Columbus's epic voyage, and the Bush administration's Moon–Mars initiative had just fizzled out. Diamandis thought governments would never have the nimbleness or stomach for risk to take on the challenge of space. Worse, they prefer to smother innovation in red tape. When Diamandis presented his idea to the FAA, they said regulations wouldn't permit passengers to be in a diving airplane with their seat belts unstrapped. It took eleven years for him to overcome the objections and offer the public a nauseating but exhilarating experience.

His most noted passenger was the iconic physicist Stephen Hawking. Diamandis had met Hawking through the X Prize Foundation, and the physicist told him of his dream to go into space. Diamandis offered a zero-gravity experience instead, and Hawking accepted on the spot. But the aircraft partner said, "Are you crazy? We're going to kill the guy!" The FAA said, "You're only licensed to fly able-bodied people." (Hawking has ALS and is confined to a wheelchair.) An exception was granted in 2007 and the scientist was liberated from his withered body for eight parabolic dives. Hawking controls very few of his muscles, but Diamandis described him as having had "a shit-eating grin" on his face.[15]

At a press conference for the flight, Hawking said, "I think the human race doesn't have a future unless it goes into space. I therefore want to encourage public interest in space."

Diamandis is utopian in his belief that galloping progress in technology will solve human problems. This is best embodied in the Singularity University, an unaccredited educational institution in Silicon Valley that he founded in 2008 with inventor and futurist Ray Kurzweil. If he didn't have so much tangible success, it would be easy to pigeonhole Diamandis as a wild-eyed dreamer. He would empathize with what Robert Goddard said after the *New York Times* had declared his goals unachievable: "Every vision is a joke until the first man accomplishes it; once achieved, it becomes commonplace."[16] In humanity's future, Diamandis foresees "nine billion human brains working together to a 'meta-intelligence,' where you can know the thoughts, feelings, and knowledge of anyone."[17]

The Transport Guru

Elon Musk wants to die on Mars.

Like Peter Diamandis, he's sure that our future is in space and that we must become an interplanetary species. He was influenced by Isaac Asimov's *Foundation* series, but his vision has a darker, dystopian slant, since it's also a hedge against threats to our survival: "An asteroid or a super volcano could destroy us, and we face risks the dinosaurs never saw: An engineered virus, inadvertent creation of a micro black hole, catastrophic global warming or some as-yet-unknown technology could spell the end of us. Humankind evolved over millions of years, but over the last 60 years, atomic weaponry created the potential to extinguish ourselves. Sooner or later we must expand life beyond this blue and green ball, or go extinct."[18]

Musk made his name and his fortune starting Internet companies, but rockets are in his blood. His father was an engineer, so he was used to getting explanations of how things worked. He built rockets as a kid, but he grew up in South Africa, where there were no premade rockets, so he went to a drugstore, bought the ingredients for rocket fuel, and stuffed them into a pipe. Thirty years later, he was building rockets efficient enough to lead the budding private space industry.

He also showed personality traits that would serve him well later and propel him into the top echelon of innovators and entrepreneurs. His mother said he had devoured the entire *Encyclopædia Britannica* by the time he was nine, and he remembered much of it. She had to check that he was getting something to eat and wearing fresh socks every day. He alienated his schoolmates by correcting their minor factual errors, and he retains that blunt intensity as an adult. Musk has a reputation as someone made of cold steel, so it's not surprising that he's the inspiration for Robert Downey Jr.'s depiction of Tony Stark in the *Iron Man* movies, the playboy inventor in a flying, weaponized suit.[19]

Musk has a lot in common with Richard Branson. Both are billionaires who like to take risks and challenge conventional wisdom. Both have made their mark in multiple transportation industries—Branson in rail, aviation, and space; Musk with electric cars, spacecraft, and his hyperloop aviation concept. They're both committed philanthropists. But it's hard to imagine Musk admitting to a debilitating weakness, as Branson did with his shyness, or getting up in drag to pay off a bet, as Branson did when he lost a bet with the boss of Air Asia and served as a stewardess on a charity flight from Perth to Kuala Lumpur in 2013.

Musk received degrees in business and physics from the University of Pennsylvania; he spent only two days in an applied physics PhD program at Stanford before leaving to pursue his entrepreneurial aspirations. He said he was just trying to pay the rent when he started a web software company with his brother to market online city guides to the newspaper industry. He sold Zip2 to Compaq for $307 million in 1999. That same year, he founded a company for online financial services and e-mail payments. He built the PayPal brand and three years later sold it to eBay for $1.5 billion. Good ideas are contagious—the creators of Yelp, YouTube, and LinkedIn all worked with him at PayPal.[20]

Musk stands out even in the overachieving cadre of space pioneers. He has the technical background to develop new technologies, he is a hands-on manager, and he has the personal wealth to fuel his dreams.

The pace of Musk's activities is breathtaking. In the consecutive years beginning in 2002, he started SpaceX, Tesla Motors, and Solar-City. The latter is the largest provider of solar power systems in the United States, riding a market that has grown by a factor of ten since 2009. SolarCity also makes charging stations for electric vehicles, helping solve a chicken-and-egg problem in which people are reluctant to buy electric cars when there isn't the infrastructure to support their use. Every Friday, Musk drives 20 miles from his 20,000-square-foot Bel Air mansion to the converted hangar south of the Hollywood Park racetrack where Tesla Motors has its research facility. He normally drives a Tesla Roadster, but his guilty pleasure is a 1967 E-Type Jaguar, which he says is like a dysfunctional but exciting girlfriend. Tesla only sold 2,500 Roadsters in its first nine years, but Musk is confident he can take on giants in the "green" car market like Toyota, which sold more than 600,000 hybrids in 2012.

Musk also commutes to the SpaceX facility, where visitors are greeted by a life-size statue of Tony Stark in his Iron Man suit. Here the company develops and manufactures the Falcon 1 and Falcon 9 rockets and the conical Dragon spacecraft (Figure 21).

An entrepreneur's life can be an extreme roller coaster. Late in 2008, Musk was close to despair. The first three SpaceX launches had failed, financing for Tesla Motors fell through, the bank behind SolarCity reneged on an agreement, and he'd just suffered a publicly acrimonious divorce. But the next day, NASA called with a billion-dollar contract offer to service the International Space Station. In 2009, SpaceX became the first privately funded company to put a satellite in Earth orbit, and in 2012 it became the first commercial company to dock with the ISS. SpaceX has a backlog of $3 billion of orders through 2017. Musk's life, however, continues to be a white-knuckle ride. In late 2013,

Figure 21. The Falcon 9 rocket is designed by Space X and built in California. Its two-stage rocket can carry 15 tons to low Earth orbit and 5 tons to geostationary transfer orbit. Space X was founded by Elon Musk, the South Africa–born inventor and investor who made his fortune as the founder of PayPal. Musk has also been an innovator in terrestrial travel with his car company Tesla Motors.

his net worth dropped by $1.3 billion after reports of weak earnings by Tesla and SolarCity, and he separated from his second wife.

We see in this progression of space entrepreneurs the march toward youth: Rutan is in his early seventies, Branson in his early sixties, Diamandis in his early fifties, and Musk in his early forties. Yet they're all connected. Rutan is designing spaceships for Branson and he won the X Prize from the organization that was founded by Diamandis, with Musk serving on the board of trustees.

These four hard-driving men get the most publicity about commercial space travel, but there are other entrepreneurs shaking up the traditional order. Bill Stone is a sixty-one-year-old engineer who started a company to look for microbial life on Europa, one of Jupiter's moons. He's also a professional caver who has led expeditions to some of the world's most inaccessible and dangerous places. Stone has seen twenty-three colleagues die in his various adventures. He's funded by NASA to test technology for a follow-up to the proposed Europa Clipper mission, which will map the icy surface of Europa and identify places where the ice is thinnest. Then another of Stone's concepts will come into play: an ice-penetrating robot that can explore the ocean beneath and search for the chemical signatures of microbial life. He thinks discovering life on Europa would be one of the most momentous events in the history of science.[21]

Stone and the "big four" are unrelated, but they might be linked genetically. Nobody has tested them for the 7R variant of the DRD_4 gene, which controls dopamine, to see if their novelty-seeking is a heritable trait. But a research group led by Scott Shane at Case Western Reserve University has studied 870 pairs of identical twins and found a clear genetic component in the likelihood of becoming an entrepreneur.[22] Neuroscientist Barbara Sahakian at the University of Cambridge, who has correlated risky or impulsive decision making with entrepreneurial success, notes: "Our findings also raise the question of whether one could enhance entrepreneurship pharmacologically."[23]

There's no single template for being a space entrepreneur. Each of these examples has a different mix of strengths and blind spots. But in an age where cynicism is widespread and many people are scared or bewildered by technology, they share an unshakable faith in its power to improve the future.

6

Beyond the Horizon

Space Is Heating Up

After years in the doldrums, space is heating up. For decades, NASA was the only option, but now it's getting hard to keep up with the activities of American private companies that are trying to create a new business model for space travel.

We're witnessing an exciting paradigm shift. NASA has always been a technocratic institution, an agency designed to manage research and development for the good of the nation. The best parallel is with the Atomic Energy Commission, which was formed in 1946 to move assets from military to civilian hands and to regulate the peaceful use of atomic energy. During the 1960s, the race to land men on the Moon established NASA as an engine of American international prestige, but that was a unique historical occurrence.[1] NASA's share of the federal budget has since shrunk from 5 percent to ½ percent. The "military-industrial complex" still has a lot of clout, but power is shifting to private and commercial organizations. Now they vie for government contracts, but eventually the government will be reduced to the levers of regulatory control, as it is with the aviation industry.

Figure 22. The first commercially developed spacecraft to dock with the International Space Station. On May 25, 2012, the SpaceX Dragon capsule approached the station and was grabbed by a robotic arm.

The first high-profile success for the private sector came in May 2012, when the Dragon spacecraft docked with the International Space Station (Figure 22). Elon Musk's gumdrop-shaped capsule is made by a company with fewer than 3,000 employees with an average age of thirty. It's a minnow compared to NASA, with 18,000 employees and 60,000 contractors and an average age of over fifty. The docking was not without incident. Dragon aborted on its first approach to the station and then had sensor problems, but eventually astronaut Don Pettit reached out 10 meters with the robot arms and grabbed it, as wild cheers erupted at NASA's Mission Control in Houston and the SpaceX control center in California. "We've got us a dragon by the tail," said Pettit. When he went through the air lock the next day to inspect the cargo, he said, "It smells like a new car."[2]

Early in 2013, and again in 2014, Dragon was back in low Earth orbit, fulfilling the first stage of a twelve-mission, $3.1 billion contract with NASA to haul cargo to the ISS. In September 2013, the business became competitive for the first time as the Orbital Sciences Corporation docked its Cygnus capsule (after an aborted attempt a few weeks earlier), unloading hundreds of pounds of valuable supplies, including chocolate for the astronauts. Orbital has its own contract to ferry supplies to the station, valued at $1.9 billion.[3]

On the same day that Orbital was ferrying candy into low Earth orbit, SpaceX made a bold move with the first launch of a beefy version of its Falcon 9 rocket. After it successfully carried up a Canadian weather and communication satellite, flight controllers tested a new method of lowering costs by reusing the rocket. But it didn't work as planned. The goal was to leave a little fuel in the rocket, reignite it, and fly it back to the launchpad. But the restarted rocket began spinning and sloshing the fuel against the side of the rocket, which starved the engines and made the rocket crash. Musk was chipper. He tweeted, "Between this flight and Grasshopper tests I think we now have all the pieces of the puzzle to bring the rocket home."[4] Grasshopper has metal struts for legs and is designed to land back on its launchpad, like a cartoon rocket. A small prototype made eight successful test flights between 2011 and 2013, and the full-size Falcon 9 version made its first test flight early in 2014. Grasshopper is designed to fly to 90 kilometers (54 miles), just short of the Kármán line, and land as accurately as a helicopter.

Other active players have promising prototypes about to emerge from skunkworks projects, so Elon Musk needs to keep innovating. He's said that when SpaceX covers its costs with satellite launches and supply runs to the Space Station, he will turn his attention to Mars.

Virgin Galactic has competition for suborbital tourist business from XCOR. The Texas-based company is developing the Lynx rocket plane, which is designed to carry a pilot and a paying passenger up 100 kilometers and down in just under half an hour. XCOR has presold nearly 300 flights for $95,000 each. Richard Branson probably isn't worried. He has three times as many signed up, with ages ranging from eleven to ninety. The sales literature points out: "SpaceShipTwo's cabin will have lots of room for zero-G fun." Virgin Galactic has the bulk of the celebrities, including Justin Bieber, Kate Winslet, Leonardo DiCaprio, and Tom Cruise. Paris Hilton mused about her prospective flight: "What if I don't come back? With the whole light-years thing, what if I come back 10,000 years later and everyone I know is dead? I'll be like 'Great. Now I have to start all over.'"[5] It's all entertaining to contemplate, but these celebrities should read the fine print carefully—the sobering truth is that the risks are real and people will likely die as the industry goes through its teething phase. Branson has a showman's hyperbole, but he was chastened by the loss of a pilot's life in 2014.

Space Adventures is the only private space company with a track record. The US-based company has a variety of space initiatives, often using hardware developed by other companies. It partnered with the Russian Space Agency to send seven civilians to the International Space Station between 2001 and 2009. The Russians suspended the arrangement due to limited Soyuz capacity but plan to start including paying passengers again in 2015, when British music superstar Sarah Brightman is scheduled to go up. She and others are paying $45 million for a two-week stay—more than the $20 million paid by the first set of space tourists but less than the $62 million the Russians charge when American astronauts hitch a ride.

Space Adventures hopes to get a piece of the suborbital action too, and they have ambitious plans for a commercial lunar flyby, starting in 2015. One unnamed person has already paid $150 million for this trip-of-a-lifetime and the company is in negotiations to sell a second seat.[6] The ubiquitous Richard Branson is also talking about Moon trips, but first he has to get an orbital launch vehicle and build a space hotel.

Robert Bigelow is sure he'll be the first to put a hotel in orbit. An iconoclastic billionaire who started the Budget Suites hotel chain, he now has higher-level accommodations in mind. Bigelow is quirky—he believes in UFOs and the power of prayer, and not in the big bang theory—but he's not to be taken lightly. His company has launched two inflatable prototypes that are still in orbit, albeit slightly deflated. NASA was impressed enough to order a unit for the Space Station, to be delivered in 2015 by a SpaceX Dragon rocket. He also plans to work with SpaceX to put a capacious 330-cubic-meter bubble in orbit. That could hold six people in relative comfort. For the smaller 110-cubic-meter version, his business model calls for $50 million to buy a return flight and a two-month stay. A year of naming rights for advertisers costs $25 million. Bigelow's products are all vaporware, so it surprised many people and was an important milestone when the company signed an $18 million contract with NASA in late 2012 to build an inflatable module for the Space Station.

The dark horse in the new space race is Blue Origin. Established by Amazon founder Jeff Bezos, Blue Origin is following an incremental approach to go from suborbital to orbital flights. The company motto is *Gradatim Ferociter*, Latin for "Step by step, ferociously." Because Amazon so deftly progressed from online bookseller to merchandising behemoth, most experts expect Blue Origin to be a major player in space. But company documents originally projected suborbital tourist flights once a week by 2010—like their competition, they've been overly optimistic. The company website hasn't announced the expected date of its first flight with paying tourists.

Billionaires Bezos and Bigelow are both notoriously publicity-shy, and there's amazingly little public information about Blue Origin. It was founded in 2000, but its existence was only revealed in 2003, when Bezos started rapidly aggregating land in Texas under a set of shell companies. Like SpaceX, Blue Origin will use a vertical takeoff and landing (VTOL) rocket that's fully reusable.

After being named valedictorian of his high school class, the eighteen-year-old Bezos said he wanted "to build space hotels, amusement parks, and colonies for two or three million people who would be in orbit."[7] Neal Stephenson, the author of *Snow Crash* and other science fiction novels, worked part-time for Blue Origin for several years.

Meanwhile, NASA isn't simply giving up and passing the baton. It's like an older brother with achievements under his belt who suddenly has a set of young, talented, and rambunctious siblings. NASA has been outsourcing much of its cargo-carrying business,[8] but it has ambitious plans that bump up against the limitations of the budget. These plans require a beefy rocket to get large payloads into Earth's orbit. It clearly chafes against agency (and national) pride to pay the Russians to ferry American astronauts to the International Space Station. The Constellation program was announced in 2005, with goals of resupplying the ISS and eventually launching manned flights to the Moon and Mars. But when a combination of technical problems, delays, and budget cuts left the Constellation program in disarray, President Barack Obama killed it in February 2010.

A few months later, the Space Launch System (SLS) rose phoenix-like from the ashes of Constellation.[9] It will reuse parts of the technology planned for Constellation and keep many of the same contractors in place—an expedient move, since the work is being done in some pivotal congressional districts. The launch vehicle will be upgraded in stages to lift 130 metric tons, making it more powerful than the mighty Saturn V. The newly designed Orion spacecraft will eventually carry six astronauts; in late 2014 it had a successful test flight.

All dressed up and nowhere to go? NASA officials are acutely aware that such an impressive and expensive capability needs a compelling destination. But the paymasters are unpersuaded by the Moon and they recoil at the cost of Mars. Here was President Obama in a major space policy speech given at the Johnson Space Center on April 15, 2010: "I understand that some believe that we should attempt a return to the Moon first, as previously planned. But I just have to say pretty bluntly here: we've been there before."[10] As a goal, NASA came up with the Asteroid Redirect Mission. This idea is to use a robotic spacecraft to pluck a small asteroid out of deep space and haul it into a stable orbit around the Moon, where it could be studied more closely.[11] NASA has a number of promising missions under development and this one was seemingly plucked out of thin air to be a centerpiece of NASA's strategy.

Advisory committees and senior planetary scientists have been skeptical of the mission, and it faces an uphill battle to be funded, let alone executed. Meanwhile, NASA's overachieving young siblings are going from strength to strength.

Bound in Red Tape

We're used to being bound to the Earth by gravity, but the nascent commercial space industry is in danger of being bound to the Earth by bureaucracy.

In 2006, the US Government released 120 pages of rules for space tourism, ranging from preflight training to medical standards for the passengers. Most of the regulations are easily followed, such as requiring those flying the spacecraft to have FAA pilot certificates and those just along for the ride to sign a form saying they had been informed of the risks involved. Other rules are vexing space entrepreneurs.[12] The most troublesome law is America's International Traffic in Arms Regulation (ITAR). Rocket systems are like tanks and guns, in that a license is required for their export. But a license is also required if they are worked on by a non-US citizen, or even shown to a non-US citizen. ITAR controls are the bane of many researchers, as they have been applied to detectors and electronic systems that have no real strategic importance. *The Economist* estimates that strict ITAR controls on satellite technology have halved the US share of the global commercial-satellite industry since 1999.[13]

Virgin Galactic has been stung by ITAR. It operates out of Spaceport America in New Mexico and has an international client list. Export regulations delayed Virgin Galactic's deal with Burt Rutan for SpaceShipTwo by several years, and Rutan doesn't mince words when talking about his dealings with the FAA: "The process just about ruined my program. It resulted in cost overruns, it increased the risk for my test pilots, did not reduce the risk to the non-involved public . . . and removed our opportunities to seek innovative safety solutions."[14] Then there's the problem of international passengers, who might not be allowed to see the insides of a spacecraft governed by ITAR. If British ticket-holders arrive at the Spaceport only to be sent home, or told they can go up wearing a blindfold, it won't be good for business. Virgin finessed the problem by designing its procedures so passengers don't see behind the scenes, but they'll face another headache when they pursue their plan to launch from Abu Dhabi in the United Arab Emirates, since the UAE isn't classified as a "friendly country" under ITAR.

The issue isn't unique to the United States. All spacefaring nations are trying to spur private investment. In Europe, Arianespace has half of the world market for satellite launches. It gets big government subsidies but is also hampered by the hyperbureaucracy of the European Union. The Russian Government has sold the majority of RCS Energia to private investors, but Russia is hostile to entrepreneurs, so Energia is locked into forty-year-old Soyuz technology. At the moment, the UK regulatory environment is so forbidding that Virgin Galactic is unable to launch from Branson's home country. However, he caught a break in May 2014 when the FAA cleared Virgin Galactic to launch into space from its Spaceport America facility.

Another issue is insurance. Space insurance is a simple extension of other kinds of travel insurance, but insurers still haven't calculated the exact risk. Rockets have significant but highly variable failure rates, and satellites are typically insured for 10 percent of their replacement cost, which can be tens of millions of dollars. For private launches, premiums are being quoted at about $300,000 for $100 million of coverage. Big rockets benefit from federal indemnification in the United States, which means losses beyond $100 million and up to $3 billion would be covered by taxpayers. Which leads to an issue that space tourism companies don't like to dwell on.

People are going to die.

Consider this passage from the *Columbia Accident Investigation Board Report*: "There is great risk in placing human beings atop a machine that stores and then burns millions of pounds of dangerous propellants. Equally risky is having humans then ride the machine back to Earth while it dissipates the orbital speed by converting the energy into heat, much like a meteor entering Earth's atmosphere."[15]

This explains the two shuttle disasters, which account for most of the twenty-one fatalities in the history of the space program (three astronauts died on the ground in the Apollo 1 fire). In 1986, an O-ring on one of Challenger's solid rocket boosters failed during the fiery ascent and led to an explosion. In 2003, a breach in a protective panel allowed the heat of Columbia's reentry to penetrate and then destroy the spacecraft.

Space travel is indeed dangerous, though not quite as dangerous as you might think. Interestingly, given its smothering influence in other areas, the FAA has been very casual with vehicle certification. For now, they've agreed to license private spacecraft without certifying, as they do for aircraft, that they are safe to carry people. To quote the regulations: "The FAA has to wait for harm to occur or almost occur before it can impose restrictions, even against foreseeable harm." So safety criteria will only be applied when specific problems arise, or there's an actual fatality rate. Meanwhile, space tourists will have to waive any claims against the American government and the operator. Which begs the question: What are the risks?

As of late 2013, about 540 people have been in space, and twenty-one have died, a fatality rate of 3.9 percent. The result is similar if counting launch or reentry attempts that have killed their crew; for both Soyuz and the Space Shuttle, which account for the vast majority of launches, the fatality rate is 2 percent.[16]

The statistics are reassuring, but the particulars of the losses are chilling. It was soft-pedaled by the media, but both of the doomed Space Shuttle crews almost certainly survived the initial incident and were conscious as they plunged to Earth. Some of the Soviet losses were

equally grim, when the details emerged from a veil of secrecy. When
Soyuz 1 crashed in 1967, cosmonaut Vladimir Komarov knew he was
going to die and raged against the engineers for ignoring prior warnings. Three cosmonauts died in 1971 returning from the Salyut 1 space
station. Their ventilation valve ruptured 100 miles up, exposing them to
the vacuum of space and asphyxiating them. There were also some close
calls. The most memorable was Apollo 13, but in 1965 the Russian Voskhod 2 spacecraft missed its reentry site and the cosmonauts landed in
a heavily forested wilderness at night. The two cosmonauts huddled in
the cold, gripping pistols as wolves and bears roamed outside. The first
Moon landing was so risky that President Richard Nixon had a speech
prepared in case Neil Armstrong and Buzz Aldrin were stranded. It
read: "Fate has ordained that the men who went to the Moon to explore
in peace will stay on the Moon to rest in peace."[17] If that had happened,
America's space program might have played out very differently.

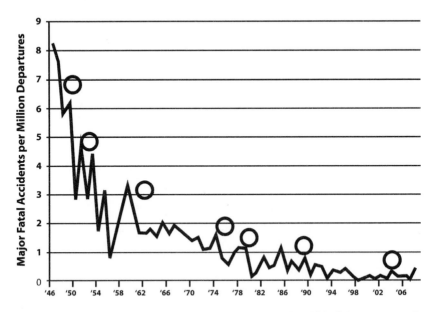

Figure 23. The accident rate in US commercial aviation since World War II. Improvements in
safety marked by circles represent (from left to right): pressurized cabins, radio communication, long-range radar, radar navigation, automation, autopilots, and large new jets.

How do these fatality rates compare to more prosaic modes of travel and risks we take without thinking about them? Commercial aircraft are remarkably safe, with 1.3 deaths per hundred million miles flown in 2008. That converts to a lifetime probability of death in an airplane of one in 20,000, or 0.005 percent. In 1938, a more pioneering era in aviation, odds of death while flying were ten times higher. But the eye-popping number is the death rate from driving, giving a lifetime probability of death of one in 84, or 1.1 percent.

So, assuming it's a once-in-a-lifetime joyride, traveling into space is four hundred times more dangerous than flying but only twice as risky as driving (Figure 23).

———————

In the early fifteenth century, the eunuch Chinese admiral Zheng He sent a fleet of 320 ships and 18,000 men on seven major voyages to India, Arabia, and Africa. Their goal: to seek out new curiosities and animals and make any civilizations they encountered submit and swear fealty to the Chinese emperor. But that vast effort was squashed. Nobody in China was allowed to own a ship and foreign trade was discouraged. Exploration simply ended. At the end of that century, Europeans began to explore in ships that were much smaller and less sophisticated than the ships of the Chinese fleet. They had some government seed money but the exploration was spurred mostly by trade and colonization. Some of these settlers embraced free enterprise and declared freedom from the smothering embrace of their former rulers, becoming the United States of America. Therein lies a lesson for the best way to go beyond the horizon into space—accept the risks and give the visionaries a free rein.

Rockets Redux

Bureaucracy is a human construct. An optimist might imagine that it can be reduced or even avoided in an ideal world. But physics is more obdurate. So the young Turks of the commercial space industry face something they can't duck: the tyranny of the rocket equation.

As we've seen, rockets are machines for generating momentum. They spew gas out of a nozzle at high speed and the rocket attached to it goes in the opposite direction. Isaac Newton defined the physics and Tsiolkovsky codified it into an equation with three variables. Specify two of the variables and the third is cast in stone. No sleight of hand or cleverness can change that fact. One variable is the energy needed to work against gravity and get to the destination, which for low Earth orbit corresponds to accelerating from rest to eight kilometers per second. The other two variables relate to the fuel: how much energy or impulse it provides, and what fraction it is of the total rocket mass.

The energy for a rocket comes from rearranging atoms. So modern rockets work at the limit of what's possible in chemistry. (One day we may be able to power rockets by rearranging atomic nuclei in a fusion reaction, but for now we're stuck with chemistry.) The most efficient reaction uses combustion, or oxidation, of hydrogen. It's clean burning, because the product is water, but it has the big complication that both molecules are gases at room temperature, so they both must be cooled and pressurized into liquid form. How does hydrogen-oxygen burning compare with other common fuels? It releases three times more energy per kilogram than gasoline or natural gas, five times more energy than coal, and ten times more energy than wood. The only thing that comes close is a highly refined and volatile kind of kerosene called RP-1.[18]

With the fuel specified, the final variable is locked into place. A solid or kerosene-powered rocket must be 95 percent fuel, and a hydrogen-oxygen–powered rocket must be 85 percent fuel. The latter number was

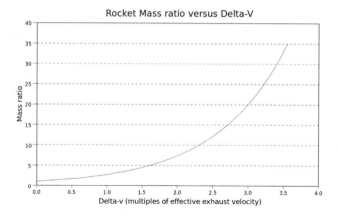

Figure 24. The rocket equation describes how the change in speed, or the acceleration, of a rocket depends on the ratio of the total initial mass, or the payload plus propellant, to the final mass, the payload. For an acceleration required to reach orbit, most of the initial mass must be propellant.

the fraction of total mass in fuel when Saturn V and the Space Shuttle launched. Just compare the 85 percent fuel fraction to that of a cargo plane (40 percent), a diesel train (7 percent), a car (4 percent), or a container ship (3 percent). These spacecraft were both mostly just hauling fuel around; the actual payload was 4 percent for the Saturn V and 1 percent for the Space Shuttle (Figure 24).

A rocket has engines, nozzles, and fuel tanks. It has a lot of plumbing that has to deliver pressurized liquid propellant swiftly and accurately to the combustion site. It has a structure that must survive the stress, vibration, and g forces of launch. It has to be able to fly in air and in the vacuum of space. Unlike the Saturn V and Space Shuttle launch systems, it must be reusable. To achieve lightness and strength, it has to be made from a combination of aluminum, titanium, magnesium, and epoxy-graphite composites. The constraints of the rocket equation are so severe that the engineering margins are very small. Testing and modeling are only taken to 20 to 30 percent above the designed limit. Imagine driving a car at 60 mph and then drifting up to 75 mph, only to have your car blow up.

Despite the structural problems that led to the loss of two Space Shut-

Figure 25. The most powerful and highest-performance rocket engine ever built. The RS-25 was built by Rocketdyne as the main engine for the Space Shuttle, generating 420,000 pounds of thrust at launch. It will also be used on the successor to the Space Shuttle.

tles, its engines were superb and had the best performance ever achieved in a rocket. The external tanks were as big as a family home. They held a combustible cryogenic fluid that was chilled to –250°C, pressurized to 60 pounds per square inch, and delivered to the engines at 1.5 tons per second. The Space Shuttle was like a Ferrari, but it was a project driven by performance, not cost. Thousands of man-hours were required to refurbish it between launches. The astronauts knew they were riding in an exquisite machine, where hundreds of thousands of parts had to work in perfect synchrony (Figure 25). They also were uncomfortably aware that it had been built by the lowest bidder. Rocket technology hasn't improved much since the 1960s. But there's a cheaper way to design rockets.

When Elon Musk thinks about rockets, he thinks like a physicist. Just as rocket fuel works by rearranging atoms, so does building a rocket. He observes that the cost of rocket materials represents about 2 percent of the cost of the final rocket. That compares to 20 to 25 percent for a car. SpaceX has innovated in ways to save on materials costs, such as making welds without riveting. They fabricate most of their components in-house, rather than pay nosebleed prices to subcontractors. They have simplified construction, with many common components among their launch systems. And they don't file patents, since Musk says the Chinese would just use them as a recipe book.[19] SpaceX is building a Toyota Corolla, not a Ferrari.

The other private space companies are following similar strategies. It's too early to tell how far they'll drive down costs, but the early results are promising.[20]

A benchmark for efficiency in space technology is the launch cost per kilogram delivered to low Earth orbit. Saturn V, the Space Shuttle, the American Delta 2, and the European Ariane 5 rocket each delivered payloads for about $10,000 per kilo. The prolific Russian Soyuz, with nearly 800 launches, costs roughly $5,300 per kilo. Its sturdy successor, the Proton-M, comes in at about $4,400 per kilo. The Chinese are tight-lipped about the economics of their Long March rocket, but they say they can't beat the cost of the Soyuz. Virgin Galactic's price tag of $250,000 per person works out to $3,000 per kilo for an average

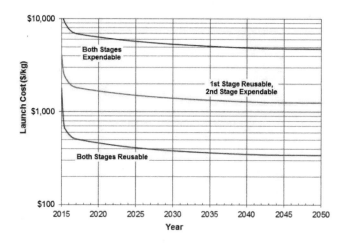

Figure 26. Projected launch costs based on current rocket technologies. Note that the vertical scale is logarithmic. Expendable rockets such as the Saturn V and the Russian Proton will always be expensive; even partially reusable technologies are difficult to force below $1,000 per kilogram. Fully reusable launch vehicles can transform the economics of space travel.

adult, but that's only suborbital. SpaceX aims to go one better. It claims its new Falcon Heavy rocket will lower the cost to low Earth orbit to $2,200 per kilo (Figure 26).

Living in Space

Being weightless is unnatural and uncomfortable. The airplane that gives people a taste of zero g is called the Vomit Comet. About half of all visitors to space experience motion sickness, which may consist of nausea, vomiting, dizziness, headaches, and general malaise. It usually wears off after two or three days, but 10 percent of all astronauts suffer a long-term and debilitating form of space sickness that interferes with their ability to function normally.[21]

This has been known since the early days, but astronauts expected to have the "right stuff" were reluctant to report it, and often they were grounded if they did. US Senator Jake Garn became the poster child for puking when he flew on the Shuttle in 1985, and astronauts since then have referred to the "Garn scale" to rate their own symptoms. Garn was essentially incapacitated during his entire flight. Oceanographer and NASA researcher Robert Stevenson said, "Jake Garn . . . made a mark in the Astronaut Corps because he represents the maximum level of space sickness that anyone can ever attain, so the mark of being totally sick and totally incompetent is one Garn. Most guys will get maybe to a tenth Garn, if that high."[22]

A Dutch researcher did experiments where people were subjected to human centrifuges (often used for astronaut training) and observed that they suffered similar symptoms after leaving the centrifuge.[23] So the trigger for space sickness is adapting to a different gravitational force, rather than the absence of gravity. Technically, astronauts are never subject to zero gravity—gravity is everywhere in the universe. But in Earth orbit, astronauts and their "container" freefall at the same rate as their forward motion, so they "float" inside the container. *Micro-gravity* is a more accurate term.

Thanks to their Mir Space Station, Russians hold the records for the longest time in zero g. Valeri Polyakov spent 438 days on Mir in the mid-1980s and three other cosmonauts spent more than a year. By going up six times, Sergei Krikalev racked up 803 days in space. How these cosmonauts fared is providing vital insights into what to look for when we send humans to Mars.

The full set of physiological effects would give any potential space-farer pause. A lack of gravity causes body shape to morph. Astronauts get 2 or 2.5 inches taller, but that extension (and subsequent compression when they return) is quite painful. Internal organs drift upward, faces get puffy, waists and legs shrink, and the result is a cartoonish image of a strongman. But this is an illusion; without gravity to fight against, muscles atrophy and bones get thin and brittle. Astronauts work out for two or three hours a day to combat these effects. The heart also weakens, and blood pressure may lower to the point where a person occasionally can pass out. Immune systems weaken in space, and the upward migration of body fluid leads to congestion and headaches. Glucose tolerance and insulin sensitivity vary wildly in space.

Since they're not shielded by the Earth's atmosphere, cosmic rays cause low-level brain damage, and they may accelerate the onset of Alzheimer's disease. Eyes suffer, too; Russian cosmonaut Valentin Lebedev suffered progressive cataracts and went blind after spending eight months in orbit. The cosmic rays also create brilliant flashes of light in the eyeball, creating a kind of private disco ball.[24]

Decent food combats physical discomfort, or at least boosts morale. Early astronauts had a fairly grim diet of flavored paste from a tube or bite-size, freeze-dried snacks. Floating crumbs and drops of liquid cause havoc with sensitive electronics, so they were avoided at all costs. Now the choice is much better. Astronauts can feast on pouches of shrimp cocktail, beef stroganoff, and cherries jubilee. Many of the technologies that NASA developed for preparing, sealing, sterilizing, and heating meals have been adopted by the food-service industry and are enjoyed by couch potatoes the world over. The Space Shuttle's fuel cells generated water, providing the side benefit that water to reconstitute the food was made in orbit, saving launch weight.

On the orbiting space stations, the scene is quite familiar. Skylab astronauts gathered around a table where they could "sit" using foot straps, and they each had a knife, a fork, and a spoon, plus a pair of scissors to cut plastic seals. The International Space Station uses an eight-day rotation of meals, which could become monotonous if you were up there a year. Half the meals are American and half are Russian, with crews getting to taste and vet the food of the other country in their training sessions. The diet of burgers and borscht has broadened as people from other countries have started to fly aboard the station.[25]

Astronauts get three meals and occasional snacks. Those having the midnight munchies must dip into the reserve supplies, which are brought in case landing is delayed for any reason. Raiding someone else's food is a major transgression; every astronaut's meals are customized according to their preferences and marked with a colored dot. Voyages have unraveled over less—the *Caine* mutiny over Captain Queeg's missing strawberries comes to mind. Getting balanced nutrition is tricky due to the limited range of food available. Vitamin D is cranked up, since there's no sunlight to help produce it and a deficiency would lead to excessive bone loss. Iron is dialed down because astronauts don't make enough red blood cells to absorb the usual amount. Calories stay about the same as on Earth.

What goes in must come out. Containing and processing waste is a real problem in the close confines of a spacecraft or space station. NASA makes astronauts use a "positional trainer" that teaches them how to guide their feces into a two-inch-diameter opening, all while watching a video shot from underneath looking up. Diapers are a thing of the past, but when the space toilet has a problem, astronauts must resort to even more primitive collection methods. In 2008, the International Space Station's lone toilet broke. TV satirist Stephen Colbert mocked NASA on his show for resorting to what they called "a bag-like collection system." When NASA held a write-in naming contest for its next-generation toilet, Colbert rallied his viewers to inundate the NASA website. He won. But NASA wimped out and declined to name the commode the "Colbert," dubbing it instead "Tranquility."[26]

For the few hundred who've been up, the experience transcends any physical discomfort. What a sublime experience it must be—to traverse in ninety minutes what took Jules Verne's Phileas Fogg eighty days, to view the curved limb where sky shades into night, to glimpse beyond the horizon. And that's just the edge of space. Beyond that, a vast array of fascinating worlds beckons.

7

A Plethora of Planets

The Pale Blue Dot

As the difficulties of astronauts make clear, we're perfectly adapted to our environment. The Earth fits us like a glove. The gravity, the length of day and night, the composition of the air, the climate—we survived and spread around this planet for tens of thousands of years without any assistance from technology.

So what does the Earth have to teach us about leaving home?

It shows us that we can adapt and survive in places where the physical environment is challenging. As we've seen early in this book, humans spread out of Africa and adapted to many inhospitable climates, such as the arid deserts of the Middle East and the frozen tundra of Siberia. Even today, descendants of these early voyagers make their homes in inclement places. Consider, for example, the durability of the people who live at the hottest, highest, driest, and coldest places on Earth.

The Timbisha tribe of Native Americans has lived near Furnace Creek in the Mojave Desert for more than a thousand years. Prospectors on their way to the California Gold Rush in the 1840s named this place Death Valley; in the summer, it can reach a scorching 134°F (57°C).

The land is harsh, but until the traditional way of life was encroached upon in the last century, it provided the Timbisha with all they needed. The tribe traveled seasonally to harvest wild fruit and seeds. Piñon pine nuts and mesquite beans were major parts of their diet, augmented by lizards and rabbits.[1]

Thousands of miles to the south, in the Peruvian Andes, indigenous people still live at an altitude of 18,000 feet (5,100 meters), high enough to give anyone who is unacclimated headaches and other symptoms of altitude sickness. Their nomadic lifestyle has been lost, replaced by backbreaking work in gold mines. At La Rinconada, miners live next to the mine and suffer mercury poisoning as a result. They receive no wages, but on the last day of every month they're allowed to take as much ore from the mine as they can carry on their shoulders. As usual, man's biggest indignities come from man and not from the land.

Continuing down the spine of the Andes, inhabitants of Chile's Atacama Desert live where some places have never recorded rainfall. Some riverbeds have been dry for more than 100,000 years, and scientists come to this stark landscape to test instruments for future missions to Mars. The oldest mummified remains predate analogous Egyptian relics by thousands of years. About 20,000 Atacameños still live here, although their Kunza language is now extinct. They herd llamas and they grow maize to eat and ferment it into moonshine. They use the chañar berry for syrup, jam, and a medicinal remedy. Some families still climb the side of the dramatically sited Licancabur volcano to make an animal sacrifice on June 21, the southern winter solstice. In this ceremony, a knife is plunged into the chest of a sheep or goat and its still-beating heart is pulled out and held up as the Sun rises.[2]

Oymyakon in Siberia is the coldest permanently inhabited place on Earth. Pen ink freezes here, metal sticks to skin, spit freezes before it hits the ground, birds fall from the sky in midflight, and coffins rise from the ground as the permafrost thaws and refreezes. The Evenki people eschew the town and prefer their nomadic lifestyle, herding reindeer.

They set up and break camp every day to follow the herd and work their sleds 20 miles or more through rugged terrain and snowdrifts. Lard is a staple of their diet. When a reindeer dies, they eat every part. The night shift consists of protecting the herd from marauding wolves while armed only with a spear carved from a birch branch. Ancient ancestors of the Evenki traveled over the Bering Strait when it was a land bridge to become the first Americans.[3]

We humans have adapted physically and genetically to harsh environments over thousands of years. Populations who have lived for many generations in arctic or subarctic climates have larger bodies than those in warm or tropical climates; the large mass means a lot of heat is generated, but the smaller surface area relative to the mass means it's radiated away inefficiently. This principle explains the difference between the Masai of East Africa, who are tall and slender with long limbs, and the Inuit of the far north, who are short and squat. The populations have analogous differences in body-fat storage and basal metabolic rate. At high altitudes, mechanisms for adaptation include raised hemoglobin production (natives of high mountain valleys in Peru and Bolivia) and increased oxygen delivery to muscles via wider arteries and capillaries (natives of Nepal and Tibet).

Without technology, the highest and lowest parts of the Earth are uninhabitable. Just 160 Western climbers have ascended the six vertical miles of Mount Everest without oxygen (compared to more than 4,000 who have had an assist from oxygen), and the depth limit for unassisted diving is 410 feet. With technology, these limits stretch to Alan Eustace and his supersonic free fall in the pressurized suit from 135,890 feet, and film director James Cameron's dive 35,790 feet to the bottom of the Mariana Trench in the DeepSea Challenger sub. These are transient experiences, putting the adventurers in mortal danger, but only for a brief time.

"All of humanity's problems stem from man's inability to sit quietly in a room alone," wrote Blaise Pascal in his treatise *Pensées* in 1669.

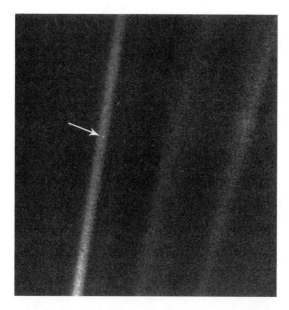

Figure 27. The Earth from a distance of four billion miles. The "Pale Blue Dot" was named by Carl Sagan after the Voyager spacecraft looked back at the Earth in 1990. If we travel in space, we may have to venture a great distance before we find a planet as habitable as ours. (The streaks are artifacts from the camera optics.)

We're wanderers. We need social interaction. We crave stimulation. All these requirements will be difficult to meet when we leave the sanctuary of the Earth. When humans go into space, they're confined, in a limited sensory environment, and cut off from normal interactions with the majority of the "tribe."

In 1990, the Voyager 1 spacecraft paused at the edge of the Solar System, twelve years after its launch and six billion miles from Earth, and took a picture looking back. The iconic image of our planet as a pale mote suspended in an ocean of black space was called the Pale Blue Dot (Figure 27).

Carl Sagan requested the image and reflected on it in his 1994 book, *Pale Blue Dot*:

Consider again that dot. That's here. That's home. That's us. On it everyone you love, everyone you know, everyone you ever heard of, every human being who ever was, lived out their lives. The aggregate of our joy and suffering, thousands of confident religions, ideologies, and economic doctrines, every hunter and forager, every hero and coward, every creator and destroyer of civilization, every king and peasant, every young couple in love, every mother and father, hopeful child, inventor and explorer, every teacher of morals, every corrupt politician, every "superstar," every "supreme leader," every saint and sinner in the history of our species lived there—on a mote of dust suspended in a sunbeam.[4]

Habitable Spots

Life on Earth has permeated every imaginable ecological niche. Humans require a "Goldilocks zone" of temperature, pressure, and atmospheric chemistry. Microbes are not so limited. Microscopic forms of life can thrive above the boiling point of water and below its freezing point, at atmospheric pressures from a tenth to hundreds of times that at sea level, and at pH values ranging from drain cleaner to battery acid. Life is found high in the stratosphere, inside rock, and at the edge of deep-sea volcanic vents.

Collectively, these microbes are called extremophiles. Biology adapts so readily to adverse conditions that extremophiles define the norm; it's the fragility of large mammals like us that's unusual.[5]

Extremophiles aren't all microbial. The tardigrade is an animal barely bigger than the head of a pin, with eight legs, a tiny brain, an intestine, and a single gonad. Colloquially referred to as "water bears," tardigrades can withstand temperatures above the boiling point of water and below less than a degree above absolute zero, pressures greater than the deepest ocean trench and the vacuum of space, and a thousand times greater radiation than other animals.[6] Perhaps their best trick is cryptobiosis. The tardigrade brings its metabolism down almost to a halt and it dries out to have less than 3 percent of its weight in water. When water is added, it reanimates.[7] The tardigrade has lessons to teach us about how to survive in space.

As humans move beyond the Earth, a key concern is habitability. We can travel in the self-contained, sealed environment of a spaceship, but the energy and materials cost of sustaining that environment is huge. It will be much easier if energy is available at the remote location and if life can be produced from raw materials extracted there.

Life on Earth is diverse but unified: Elephants, butterflies, and fungal spores all share the same genetic code and all emerged from a single common ancestor about four billion years ago. The expression of that genetic code creates an astonishing array of life forms and functions. Earth is the only place we know of with life. With only one example of biology to study, scientists can't specify the full range of habitats for life. Life evolved by natural selection to almost "fill the envelope" of physical conditions here on Earth, so it's tempting to think that life elsewhere will occupy the full range of physical conditions on an alien planet. But that's an assumption; until we know how life started on Earth or find another example of life beyond Earth, it's possible that biology was a fluke. The burgeoning subject of astrobiology tries to understand how life began on Earth, what the sites for life elsewhere are, and whether or not any of them actually host life. Without a "general theory" of biology to set expectations, astrobiology is largely an empirical subject. To know whether or not we're alone in the universe, we have to look.

The minimum requirements for biology as we know it are carbon, water, and energy. Carbon is the basic building block of complex molecules, and organic chemistry depends on the versatility of the carbon bonds. Water is a good medium for fostering chemical reactions and building complexity, and it's a major part of all terrestrial creatures— from 40 percent for beetles to 99 percent for jellyfish. Both carbon and water are cosmically abundant, so setting them as prerequisites for life isn't very restrictive. Then there is energy. Humans are at the top of a food web reliant on the Sun, but it doesn't follow that life needs a star. Some terrestrial microbes are sustained by heat from volcanic vents or natural radioactive decay in rocks. In both of those cases, the source of energy is geological.

Habitability means something quite different for microbes and men. Microbes simply need a niche with basic organic material, some water, and a local energy source. Large mammals are fussier. They need a temperate, stable climate, which requires a specific planetary spin and orbit, since daily and seasonal variation can't be too great. The need for a steady supply of water is paramount, since the metabolic processes are regulated in an aqueous solution. Traditionally the habitable zone in astrobiology is defined as the range of distances from a star where water can be liquid on the surface of a terrestrial planet.[8]

We could visit most parts of the Solar System in the next few decades, but we need destinations where conditions aren't too harsh. Within our cosmic backyard—the Solar System—what's habitable and what's not?

The Moon and Mercury are too small to retain an atmosphere and are geologically dead. With surfaces pulverized by meteors and irradiated by cosmic rays, they are considered uninhabitable, even by microbes. Venus is a close twin to the Earth in mass and size, but volcanism in its distant past pumped so much carbon dioxide into the atmosphere that a runaway greenhouse effect occurred. The resulting atmosphere is a hundred times denser than the air we breathe. It's hot enough to melt lead and is laced with toxic ingredients such as acetylene and sulfuric acid. The verdict: nasty and lifeless.

Looking out from the Sun, we arrive at Mars. The red planet is beyond the edge of the traditional habitable zone, so too cold to host surface water, and its atmosphere is so thin that a cup of water placed on the surface would evaporate in seconds. But there's indirect evidence

Figure 28. Europa is a large moon of Jupiter far outside the conventional habitable zone, yet it contains all the ingredients for life. Under the ice pack shown here lies a kilometers-deep ocean, with heat flowing into it from the rocky interior of the moon.

for subsurface aquifers, where water can be kept liquid by radioactive heating from rock below and pressure from rock above. Mars may well have microbial life in these subterranean oases.[9] For this reason, it's a compelling target for future probes and rovers.

The gas giants were long considered completely dead. Jupiter, Saturn, Uranus, and Neptune are far beyond the habitable zone, from five to forty times the distance of the Earth from the Sun. In the 1980s, the Voyager spacecraft provided a surprise when it found Jupiter's moon Europa to be a world completely covered by oceans and ice (Figure 28). More recently, the Cassini spacecraft provided evocative details of Saturn's large moon Titan, which has large bodies of liquid and river deltas, clouds, and a thick nitrogen atmosphere. Titan is eerily Earth-like, but it's alien in its chemistry, with lakes made of ethane, methane, and ammonia. It was even more surprising when Cassini saw geysers shooting ice crystals from the surface of Enceladus. This tiny moon— no bigger than Rhode Island—has underground bodies of water, so it has all the ingredients needed for life. The best guess is that there are about a dozen moons in the outer Solar System with habitable "spots" for microbes if not for larger forms of life.[10]

We know so much about the Solar System that it frames our thinking about life beyond Earth. Our planet is peerless as a habitable world, but there are definite prospects of some forms of life beyond the "Goldilocks zone."

Worlds Beyond

If we commit to develop the technologies of space travel, then one day we'll have the capability to travel to the stars. Whether we "outgrow" the Solar System or are simply curious about worlds beyond, we'll leave the safe harbor of our planetary system and venture into deep space. Between stars is the almost perfect vacuum with typically just one atom in a sugar cube volume, 30 thousand trillion times less dense than the air we breathe. It's at a temperature of −454°F, just a whisker above absolute cold. Until a few decades ago, we could only speculate about other safe harbors. Now we know they exist.

The telescope was small, less than two meters in diameter, not large enough to crack the list of the top fifty largest telescopes in the world. The site was mediocre, not high enough to have sharp images and not far enough from the city of Geneva to be truly dark. The project was protean, a survey of binary stars to diagnose their properties by how their light changed when they eclipsed each other.

But when Michel Mayor and Didier Queloz looked at the light curve of the bright star 51 Peg, they were taken aback. The star was not part of a binary system. Instead, it had a planetary companion about half the mass of Jupiter whipping around it on an orbit just over four days long. This gas-giant planet went around its Sun-like star twenty times faster than Mercury orbits the Sun. After centuries of speculation and decades of searching, the Swiss team had discovered the first planet outside the Solar System,[11] an achievement likely to earn them a future Nobel Prize.

Their 1995 discovery kicked off a new field of science. Since then, the study of extra-solar planets, or exoplanets, has exploded.[12]

Exoplanet detection pushes the limits of technology. Jupiter reflects a hundred millionth of the light of the Sun, so a remote Jupiter will look like a feeble dot of light nestled close to a vastly brighter star. Direct imaging of exoplanets is so difficult that it only succeeded in the last decade. Mayor and Queloz used an indirect method, where the planet is unseen but is detected by its periodic tug on the central star. When a planet orbits a star, the star isn't stationary, but both orbit a common center of gravity. For example, as seen from a remote location, Jupiter makes the Sun pirouette around its edge every twelve years, which is the orbital period of Jupiter. The planet causes an oscillating motion of the star, which manifests as a Doppler shift of its light. High-resolution spectroscopy teases out this very weak signal, which is a wavelength shift of one part in ten million.[13] The Doppler method gives the mass of the exoplanet and the orbital period, which, via Kepler's law, gives the distance of the planet from its star and thus its temperature.

The discovery by Mayor and Queloz was surprising because gas-giant planets had been thought to lie far from their stars, with orbits lasting decades. Other planet-hunters assumed they would need to gather years of data before seeing a planet's signature. No one understood how a massive planet could form so close to a star.

Since 1995, the number of planet-hunters has grown. They have honed their techniques so that the detection limit has advanced from Jupiter-mass to Neptune-mass and now close to Earth-mass. Roughly one in six Sun-like stars has a planet around it, and many have more than one planet.[14] For example, the Sun-like star HD 10180, which is 130 light years from Earth, has seven confirmed planets and two more unconfirmed planets, making it as heavily populated as our Solar System (Figure 29).

The first discoveries of "hot Jupiters" were puzzling and indicated that the Copernican principle might be violated. What if the arrangement of our Solar System—small, rocky planets close in and large gassy planets farther out—was not typical? Theorists couldn't figure out any way to form a giant planet very close to a star; there simply isn't enough gas. The answer was that planets can move around. Gravity keeps planets circling the Sun, but it also subjects them to subtle forces that can make their orbits unstable, rearrange them, send them closer to the star, and even eject them from the system. Hot Jupiters like the one found by Mayor and Queloz formed at larger distances and migrated inward, parking on tight, tidally locked orbits. Gassy planets farther from their parent stars have been discovered, and there's at least one terrestrial planet for every giant planet. There are likely to be many free-floating planets, called "nomads," in interstellar space. By late 2014, the number of exoplanets was approaching two thousand.

In the past decade, a second method has been used to find exoplanets. If a system is oriented so the orbital plane is close to the line of sight, an exoplanet can transit in front of its star and cause a partial eclipse or temporary dimming of the star. The star dims by a fraction

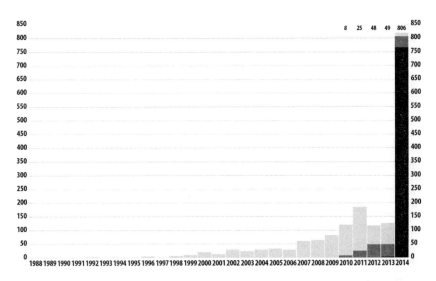

Figure 29. The number of exoplanets has surged recently with the work of NASA's Kepler telescope. The pale gray represents discoveries using the Doppler method, and medium and dark gray represent singly and multiply confirmed transit detections with Kepler.

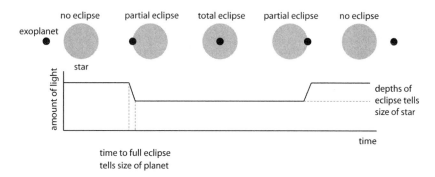

Figure 30. Most of the habitable exoplanets known have been discovered by the transit method, where the exoplanet partially eclipses and dims the parent star. Planets smaller than the Earth can be discovered with this technique from space.

equal to the ratio of the area of the planet to the area of the star; this is 1 percent for a Jupiter and 0.01 percent for an Earth crossing the face of a Sun-like star (Figure 30).

As the exoplanet count grows, the goal has shifted from finding exoplanets to characterizing them. The Doppler method gives mass and a transit gives size, so combining the two observations yields a mean density. That has been used to distinguish between gassy and rocky planets. Interpreting a single value of density can be ambiguous, but nature is imaginative enough to have made some planets that are mostly metal, some that are mostly rock, some that are mostly carbon, and some that are mostly water or ice. Evidence suggests that within this diversity are some planets that are just like home.

Hunting Earth Clones

The architect of NASA's Kepler spacecraft has called it "the most boring mission ever." The telescope mirror is one meter in diameter, the size of a coffee table and smaller than mirrors some amateur astronomers use. The telescope has been staring at 145,000 stars in a single patch of sky and measuring their brightness every six seconds.

This "boring" mission has come close to finding a direct analog of the Pale Blue Dot, a planet we might call Earth 2.0.

Kepler's goal is a census of Earth-like planets in nearby regions of the Milky Way galaxy. Its strategy is a mixture of exquisite accuracy and brute force.[15] The accuracy is required because an Earth-like planet passing in front of a Sun-like star only dims it momentarily by 0.01 percent. Reaching this level of accuracy is difficult, since the stars in Kepler's small field of view are a hundred times fainter than any star that can be seen with the naked eye. Kepler must see a transit recur several times to be sure the light dip isn't just a glitch or noise in the detector. The brute force comes in because planetary systems are randomly oriented and only a small fraction of them will be oriented so that an eclipse is visible. Odds are 1 in 215 for an Earth orbiting a Sun. If Earths exist in 10 percent of all planetary systems, then 100,000 stars must be monitored to detect a few dozen Earths. It's the proverbial needle in a haystack.

Kepler was launched in 2009 and quickly started to detect Earth-size planets, even though its sensitivity to light was a little worse than the design goals. The easiest exoplanets to detect are large ones on rapid orbits, since they cause bigger and more rapidly recurring eclipses, with a higher probability of being observed. The same is true in Doppler detection. Kepler detected a number of hot Jupiters in its first few months of operation. But as the mission progressed, smaller planets on larger orbits were detected. In 2013, the mission suffered a mortal wound when it lost the second of four reaction wheels that keep the spacecraft locked on target. It's a bittersweet ending to a fabulously successful mission, but scientists will continue to pore over the four years of data and extract evidence for Earth-like planets near the limit of detection for several more years.[16] As of April 2014, Kepler had a haul of 1,770 confirmed exoplanets and 2,400 candidates, almost all of which are likely to be confirmed.[17] Most of these exoplanets are super-Earths or larger, but some are smaller than the Earth.

Among half a million stars in Kepler's field of view, scientists focus on the third of them that are similar to the Sun. In terms of habitability, stars more massive than the Sun are poor targets because they're variable and emit lots of high energy radiation, and they live short enough lives that complex life on a nearby planet might not ever have enough time to evolve. At the other end of the mass spectrum are red dwarfs, which outnumber Sun-like stars by a large factor. Red dwarfs have slim habitable zones, so the odds of a planet being located there are low, but that is offset by the amount of such zones. In other words, the red dwarf habitable zone is small but there are many of them. When the calculation is done carefully, it turns out that there's more habitable "real estate" associated with dim red stars than with stars like the Sun. Astronomers have started doing transit surveys of dwarf stars three to ten times less massive than the Sun.

Kepler data have been used to project the total number of exoplanets in our galaxy. There are roughly 40 billion Earth-size planets orbiting in the habitable zones of Sun-like stars and red dwarf stars, with 25 percent of them orbiting Sun-like stars. That abundance means that, statistically, the nearest such planet is likely to be only twelve light years away.[18]

While looking for a "Goldilocks" situation where conditions are just right for biology, astronomers have discovered a freak-show assortment of exoplanets. Methuselah, an exoplanet 12,400 light years away, is three times older than the Earth. Since it was formed within a billion years of the big bang, it's surprising that stars had made enough heavy elements and "grit" to form a planet. The star 55 Cancri has a super-Earth so hot and dense that a third of the surface is made of carbon crushed to a diamondlike state, worth a cool 3×10^{30} if it could be brought back to the Earth. GJ 504b is a Jupiter that's farther from its star than Neptune is from the Sun. Even though it's in the deep freeze, it glows a ruddy pink color because it's shrinking due to gravity. At the other extreme, there's a planet that orbits in darkness around a pulsar, whipping around the stellar corpse every two hours. TrES-2B is a mysteriously dark planet, blacker than coal or ink, and it's not known what chemicals in its atmosphere cause it to absorb 99 percent of the light falling on it. GJ 1214b is a water world that's completely swaddled in oceans tens of times deeper than those on Earth. Finally, Wasp 18b is falling onto its star as its orbit degrades. It will enter its final death spiral in just a million years—the blink of an eye in cosmic time.[19]

Habitability depends primarily on the distance of a planet from its star. But it also depends on added heating from any greenhouse gases like carbon dioxide and methane in the atmosphere. It may also relate to plate tectonics, since the dynamism of geological activity was probably a driver for biology on the Earth. In the early oceans of the Earth, the chemical activity driven by plate tectonics is thought to be important for sustaining biochemical reactions. Modest orbital eccentricity and

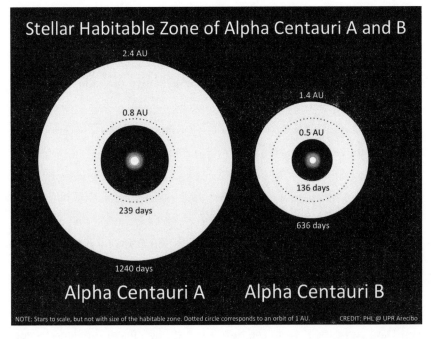

Figure 31. The habitable zones of Alpha Centauri A and B, shown as the pale wide rings. The exoplanet found around B is too close to the star to be habitable, but a habitable world might still be found. Earth's orbit is shown as a dashed circle for reference.

tilt are required to avoid big seasonal variations. However, these bounds are loose enough to accommodate super-Earths and pint-size planets or moons where conditions might be right for life.

Intriguingly, the best prospect for finding Earth 2.0 may be around the nearest star to the Sun.

In 2012, researchers at the European Southern Observatory caused a stir when they claimed the detection of a planet 20 percent more massive than the Earth orbiting Alpha Centauri B, the closest Sun-like star to us at a distance of only 4.37 light years (Figure 31). The team that found the planet was part of the Mayor and Queloz group that discovered 51 Peg, the first exoplanet, in 1995, and they made the discovery while pushing the limits of the Doppler method, trying to measure a motion of half a meter per second instead of 50 meters per second for 51 Peg. Other groups have questioned the detection, so it remains in doubt.[20]

But even if the Alpha Centauri B planet is real, this Earth-like planet isn't habitable. It's twenty-five times closer to its dim red star than the Earth is to the Sun, traveling on a three-day orbit that's probably tidally locked. More distant Earth-like planets at more temperate locations aren't ruled out, but they're below the detection limit of the best planet-hunting gear available. In another few years, the tools will be sophisticated enough to detect them. If there's an Earth so close, we will put an enormous effort into characterizing it and seeing whether it could host life. And if it's habitable, won't it be tempting to take the next logical step by sending robotic probes and then people to explore it?

PART III

FUTURE

My heart is palpitating and my skin is clammy. It's all I can do not to bolt, but of course there's nowhere to run. Josefina rests her hand gently on my shoulder. I take a deep breath and steady myself.

Minutes earlier we'd ziplined over from the station to Ark 1. I watched the Earth slide under my feet like I was skating on blue-white ice. As the ark approached, there was no sense of scale; its surface was black and seamless, reflecting no light. There would be no Hub in our next home, and no recreation—the ark was a high-tech sarcophagus.

Going through the air lock, I get a sense of the mass of this monolith. Its beryllium and carbyne laminate is designed to quench cosmic rays and stop meteorites up to the size of a pea. Inside, there are soft lights and soothing bot voices guiding us around and explaining the guidance and life-support systems, but all I can think of is the narrow long corridor shrinking to a vanishing point in front of me and the stacked translucent boxes on either side as I float by. Frosties, the people down below call us. We'll be taken to the edge of death, left there for a century, then reeled back into consciousness to explore a new world.

It's quixotic, even preposterous. But that isn't what gives me a panic attack. It's the uncompromising vessel. The ark is utilitarian, Spartan, with no grace notes. It's designed with only one thing in mind: to shield its human cargo from the indignities of space. To close on a more positive note, the bot guides us to the one common area where we'll be able to eat and rest before taking pods down to the surface.

I want to be with Josefina but the Overseers are calling the shots. Ark assignments are done by lottery. She was picked for number one and I got number two. Keeping focus in the station is harder and harder. All the tasks they design to keep us busy seem meaningless. There are more defections and ejections. The churn is so great that we speculate that there might be a second, "shadow" Academy on a neighboring lake in Switzerland. How appropriate that the Overseers would plan a larger pool so that natural selection on the station will increase the chances of mission success.

Nina. Pinta. Santa Maria. From an earlier age, small ships setting out on a vast ocean to an unknown fate.

We spend a last precious hour in the Hub on her last day. Ark 1 is completing deployment of its solar sail. The gossamer-thin membrane has unfurled on all sides to span a square kilometer; the spaceship is dwarfed by it like a stick of charcoal on a silver carpet. The sail will harness the solar breeze to accelerate the ark to the edge of the Solar System and then pulsed fusion of hydrogen atoms snatched from space will propel it to its destination. Not good-bye, she says, it's au revoir, but we both cry.

Ark 1 departs the next morning. I keep busy with my studies. There's plenty to learn to be ready to handle the rigors of the pioneer life. I stay fit. I keep mostly to myself. I'm focused. So maybe I'm not paying attention one evening at dinner when the voice over the PA says that there's been an incident on Ark 1. That a design flaw has eluded the corrective capability of the neural net. That the life-support protocol has been compromised. That it's a freak occurrence, one never seen in the sims. That in the judgment of the project office Ark 1 has been lost with all hands.

Numb. I stay that way for weeks. I've no idea why the Overseers don't pull me. They have every reason to. Maybe many others are in nearly as bad a shape. Gradually I bottom out and join in training and social activities. I become resolute. There's nowhere to go but up. After all, that's why I'm here.

Ark 2 is poised for launch. Ark 3 will follow in a few months. Its solar sail is transparent, so sunlight streams through it untouched. With one command, a weak electric current can be applied, the polarization in the thin film will shift, the material will become opaque, and Newton's laws—which killed my father—will push me away from home. As we wait for our turn in the air lock to leave the station, we all pass through anxiety into a kind of delirious anticipation. We laugh and chatter, giddy with excitement.

I have a tinge of panic as the lid clicks shut. Then I watch with mild interest as clamps cinch my wrists and ankles. The needle drops down and I feel very little pain when the IV starts to replace my blood with glycerol. As the nitrogen vents open and the lid frosts white, I keep a single thought in view: I exist.

✦

8

The Next Space Race

China Makes Its Move

Wan Hu would have been proud. Late in 2013, China sent a rocket to the Moon carrying the Jade Rabbit probe to place in the Bay of Rainbows. Despite the lyrical names, Jade Rabbit ("Yutu") is a protean and workmanlike six-wheeled rover and the Bay of Rainbows is an arid volcanic plain (Figure 32). The mission is a major landmark for the world's newest superpower; it's been thirty-seven years since any country soft-landed a probe on the Moon.

The first space race was spawned by a rivalry that cast its shadow over most of the twentieth century, cleaving the world into capitalism and communism and pitting free markets against command-and-control economies. Orbital flight was the by-product of a ruinously expensive arms buildup that took the world to the brink of nuclear war. Surely the days of space travel orchestrated by governments and colored by military aspirations are over?

Not yet, and we may be witnessing a new space race.

Here's a snapshot of current international space activity. Spending on government space programs dropped in 2013, for the first time in

Figure 32. China is the third nation to land a wheeled rover on the Moon, after the Soviet Union and the United States. The Jade Rabbit, or Yutu, rover reached the Moon in December 2013. It has a plutonium-powered nuclear reactor.

two decades, as budget cuts in the United States offset rising investment by emerging space powers.[1] The United States still has just over half of the total of $72 billion annual spending, but it has declined by 20 percent from its peak of $47.5 billion in 2009. Until relatively recently, Russia has been using buoyant oil revenues to pump lots of extra money into its space program. It's the only other country spending more than $10 billion. China is in eighth place but moving up fast; relative to GDP, its spending is still modest. That means it has the capacity to make even greater strides in space.

China is a rapidly emerging superpower, with a GDP per capita (scaled to purchasing-power parity) that will exceed that of the United States in about five years. The purchasing-power-parity comparison makes sense, since goods and services are cheaper in China and they get a lot more bang for their yuan. Chinese spending in space matches the growth rate of the economy, which has been averaging 10 percent per year for the past two decades. It makes a dramatic contrast with

Europe and the United States, where inflation-corrected spending has been flat or declining over the past two decades.[2]

China's rank of eighth in the space league is misleading. As you know from your car, objects in the rearview mirror may be closer than they appear. Let's see how they got to where they are today.

The father of the Chinese space program is Qian Xuesen. He left Shanghai to study at MIT at the same time that Mao Zedong began the Long March, a bloody retreat from the Nationalist forces that helped cement his grip on the Communist Party. Qian then worked at Caltech, where he helped famed rocket scientist Theodore von Kármán found the Jet Propulsion Laboratory. At the end of World War II, Qian and von Kármán went to Germany and helped coordinate "Operation Paperclip," which brought Wernher von Braun and other Nazi rocket experts to the United States. Qian became the foremost theorist on rocket propulsion in the country.

Then seismic forces of politics intervened. In 1950, Korea became a bloody battleground, with the United Nations and the United States supporting the South and China and the Soviet Union supporting the North. Mao felt that the world's superpowers didn't respect him, and he was convinced that only a nuclear deterrent would guarantee the security of the new People's Republic of China. Meanwhile, the "Red Scare" was sweeping the United States; Qian Xuesen was stripped of his security clearance and placed under house arrest. In 1955, he was allowed to leave the country in exchange for American pilots captured during the Korean War. Mao was delighted. He had once lamented that his country couldn't launch a potato into space, so he welcomed Qian back as a hero and put him in charge of China's ballistic missile program.[3]

China was a poor country and progress was very slow. For a while, China benefited from Soviet expertise and hardware, but in 1960, Mao accused the Soviets of backsliding on communism and of ideological impurity. So China chose to go it alone—just when the turbulence and uncertainty of the Cultural Revolution was damaging to all scientific and technical activities. As a result, China launched its first guided missile in 1966, twenty years after the United States, and its first satellite in 1970, twenty-three years after the Soviets launched Sputnik. Then came major setbacks in the mid-1990s. In 1995, a Long March 2E rocket exploded shortly after launch, killing six and injuring twenty-three. A year later, a Long March 3B rocket blew up twenty-two seconds after launch and crashed into a nearby village, with a death toll of more than two hundred.

But years of patient, well-funded work paid off. Half a millennium after Wan Hu's minions lit the forty-seven rockets that probably incinerated the Ming Dynasty official, China became the third nation to launch people into orbit using its own vehicle. Yang Liwei was dubbed a *taikonaut*—a purely invented English word designed to give China's spacefarers an equal footing with America's astronauts and Russia's cosmonauts. Since then it's been a steady upward arc. By the end of 2013, ten taikonauts had orbited the Earth in five launches. Between 2008 and 2012, China launched an average of twenty spacecraft a year. Most are doing mundane but essential work carrying telecommunications satellites into orbit.

The Chinese are acutely aware of their place in history and the way others perceive them. They also value ceremonial landmarks. So state media loudly trumpeted the launch of Shenzhou 10 in 2013, ten years after the first Chinese man traveled to Earth orbit. The crew included the second female taikonaut, Wang Yaping. She broadcast a live physics lesson to Chinese schoolchildren, joking, "We haven't seen any UFOs." As PR for the space program, it rivaled the efforts of Canadian astronaut Chris Hadfield, who played guitar and sang David Bowie's "Space Oddity" on the International Space Station.[4] Shenzhou 10 also tested the docking capabilities of the spacecraft with a module that's a precursor to China's own full-size space station.

Wang Yaping has a light touch but the general tone of the Chinese taikonaut corps is earnest and patriotic. At the Shenzhou 10 launch, President Xi Jinping told the crew: "You make all the Chinese people feel proud. Your mission is both glorious and sacred." Commander Xie Haisheng responded in kind: "We will certainly obey orders, comply with commands, be steady and calm, work with utmost care, and perfectly complete the Shenzhou 10 mission." Wang Yaping was on message as well, saying that during parachute exercises with the Air Force, "We girls all cried while singing an inspiring song 'A Hero Never Dies' on our way back after the training."[5]

It's not all been smooth sailing, however. The Chang'e 3 lunar probe was the first Chinese soft landing on an extraterrestrial body, but the Jade Rabbit rover only traveled 100 meters across the Moon before it was immobilized by mechanical failure. Engineers had not anticipated the demands of the harsh, fourteen-day-long lunar nights, and they declared that the problem was electrical not mechanical, with components suffering from "frostbite." As of late 2014, the rover was sending back limited data and was alive, but only barely.

Meanwhile, China has sent a rocket to the Moon and will bring back a lunar sample in 2017. China is also drawing up plans for a manned Moon-launch rocket that will be more powerful than the Saturn V.[6] By 2020, China could launch its own space station, just as the International Space Station is being decommissioned and crashing into the ocean. By then, it could also be in a position to land taikonauts on the Moon, half a century after the Americans abandoned such efforts. The Chinese took advantage of not having to develop a lot of space technology first. Russia was cash-strapped in the 1990s and sold its technology to the Chinese, who reverse engineered and copied it. As a result, Shenzhou looks like the Soyuz capsule and Jade Rabbit looks like the Lunokhod rover. But now the Chinese are innovating and vaulting ahead. Their Long March rocket is original and has quickly eclipsed Russian rockets.

The average age of employees in the Chinese space program is twenty-seven, less than half the age of NASA employees. A decade from now, when that youthful energy is matched by experience, China will be a formidable player.

None of this makes officials in the United States very happy.

Suspicions of Chinese motives in space run deep in American political circles. NASA officials are barred from working with Chinese nationals and Congress has barred Chinese from visiting NASA facilities without a special waiver. The ban extends to the International Space Station, even though many of the partners in that project would like to draw China into collaboration rather than treat it like an adversary. US lawmakers were incensed by a 2007 Chinese anti-satellite test that created the largest space debris cloud in history when it pulverized one of its own aging satellites. Ironically, the ITAR legislation that so frustrates American space entrepreneurs was designed to stop "unfriendly" countries from acquiring technology with military applications, but hasn't slowed the Chinese at all. Most of what they want, they build; what they can't build, they buy from other countries.

The Chinese space program is ascendant in part because they have followed the example set by the United States in the 1960s: healthy funding and a single-minded purpose. In addition, because China is a tightly controlled society run by a government with limited accountability, the People's Liberation Army is able to have a huge influence on the space program. Since its controversial 2007 test, China has continued to develop its antisatellite capabilities, and it's also within reach of having its own network of GPS satellites, which could have military as well as civilian uses. In April 2014, President Xi ordered his air force to speed up the integration of air and space capabilities.[7] China is even developing its own spaceplane, a "black project" called Shenlong, or "Divine Dragon."

For those who worry about the militarization of space, this all brings to mind the purported Chinese curse: "May you live in interesting times."

Edge of the Law

Dennis M. Hope is Overlord of the Solar System.

The sixty-five-year-old Nevada resident has used celestial property as his sole source of income since 1995. He's sold 600 million acres on the Moon, 300 million acres on Mars, and a combined 120 million acres on Mercury, Venus, and Io. He's had customers in 193 countries, with the youngest a newborn and the oldest ninety-seven. The Hilton and Marriott hotel chains have bought plots of space land. The biggest parcel is continent-size, for a cost of $13,331,000 (none sold yet), but he's moved a lot of 2,000-acre plots. An acre is a steal at $19.99 (plus $1.51 "lunar tax" and $10 for shipping and handling of the ownership certificate). He's had more than half a million customers.[8]

Surely the man is a charlatan and selling extraterrestrial real estate is forbidden by international law? Well . . . not exactly. Hope was going through a divorce and needed money. Looking out the window at the Moon, he suddenly thought, *That's a lot of vacant property.* At the library, he read Article II of the 1967 Outer Space Treaty and interpreted it as saying that no nation has sovereignty or control over any of the satellite bodies. Since the treaty made no mention of property rights of individuals, he saw his opening. He claimed ownership of all the planets and moons in the Solar System and sent the United Nations a note saying that he intended to subdivide them and sell off lots. Hearing nothing back, he set up his business.[9]

So what is the legal status of objects in space?

In 1958, the same year that NASA was formed by an act of Congress, the United Nations formed the Committee on the Peaceful Uses of Outer Space to oversee subsequent agreements.[10] The cornerstone of space law is the Outer Space Treaty, which came into force in 1967. It's been signed by more than 100 countries, including all the major players in space exploration. The treaty was triggered by the tensions of the

Cold War and it bars countries from putting nuclear weapons or other weapons of mass destruction into Earth orbit, into deep space, or on the Moon or any other celestial body. It also forbids any government from claiming ownership or jurisdiction over the Moon or any other celestial body. Countries own anything they put into orbit or launch into space, but they're responsible for any damages caused by those objects. The utopian ideal encapsulated by the Outer Space Treaty is that space exists for the "common heritage of mankind."

The follow-up to the Outer Space Treaty was the Moon Treaty of 1979. It's been signed by only fifteen countries and, tellingly, the list doesn't include any of the countries that have either been to the Moon or are developing the capability to go there.

Part of the reason the United States never signed the Moon Treaty is that Article XI says that the Moon and its resources are not subject to any sort of sovereign or private property claims.[11] The treaty requires resource extraction and allocation to be governed by a vaguely defined "international regime."

The issue has become relevant recently with the NASA plan to lasso an asteroid and put it into a lunar orbit so that it can be mined. Under the Outer Space Treaty, it's unlikely another country could put a claim on the US rock, but what if it breaks up or becomes a hazard? The liability issue has never been tested and the treaty is out of date and unequipped to deal with future space scenarios.[12]

US space policy has been tugged in two opposing directions over the past half century.

One train of thought extends into space the concept of Manifest Destiny and the ethos that led to the exploration of the American West. In fact, a frontier metaphor has framed much discourse about space policy. An advocacy group called High Frontier is explicit in its comparison; it recommends that the US Government apply nineteenth-century homesteading law and the Jamestown settlement model to colonies on the Moon. The group's director said of the Outer Space Treaty: "The

United Nations is just playing King George at the time of the American Revolution, thinking they can tell everyone else what to do."[13] Following this analogy, colonists on the Moon or Mars may start out under UN jurisdiction, but they would eventually rebel, just as the colonists rebelled against being governed by English Privy Law.

Even though America no longer dominates space exploration, former NASA administrator Michael Griffin said that when more people live off-Earth than live on it, ". . . we want their culture to be Western, because Western Civilization is the best we've seen so far in human history." This is a jaw-dropping neocolonial statement to come from the mouth of such a high-ranking government official. He may have been unaware of a prior rebuttal by Mahatma Gandhi; when asked what he thought of Western Civilization, he's reported to have said, "I think it would be a good idea." One year earlier, X Prize Foundation Chairman Peter Diamandis said, "The Solar System is like a giant grocery store. It has everything we could possibly want. . . . The Solar System's seemingly limitless energy and mineral resources will solve Earth's resource shortages."[14] This is acquisitiveness dressed as utilitarianism—if it's there and we want it, we'll take it.

The second theory of settling space uses a wilderness metaphor, where we apply values of environmental protection and preservation to space exploration. Space entrepreneurs and venture capitalists argue that turning space into a national park–type zone will squelch commercial space ventures before they get going. Until commerce spreads into space, the legal issues are hypothetical and unresolved, but the first commercial spaceflights will intensify the debates. Opportunist Dennis Hope was a ventriloquist before he owned the Moon, and he says his dummy taught him a valuable lesson: You can say anything you want as long as you're smiling.

Stairway to Heaven

Earth orbit is an excellent staging point for further exploration. The International Space Station may be unloved, but the idea isn't ill-conceived. Zero gravity facilitates the manipulation and assembly of large pieces of hardware such as rockets and habitats. The major energy cost is in struggling nearly 400 kilometers from terra firma to low Earth orbit. Going from Earth orbit to the surface of the Moon is an additional 75 percent. Going all the way to the surface of Mars only doubles the energy cost, but it's at least 140 times farther away than the Moon.

Elon Musk and other space entrepreneurs are tweaking and optimizing the rocket equation. What if it could be rendered obsolete?

Space elevators promise to do just that. Rockets are complicated, dangerous, and inefficient. Whether the fuel is solid or liquid, a rocket launch is a barely controlled explosion. The failure of a weld or a valve or a switch can spell disaster. At the major Russian launch sites, they don't even do countdowns. They stand at what they hope is a safe distance and wait for the outcome. Rocketry is based on rearranging electrons in atoms and molecules, which is a chemical energy source only three times more efficient than the venerable internal combustion engine and five times more efficient than burning coal in a fireplace. It costs at least $10 million to put anything into Earth orbit. Gravity smiles mockingly at our efforts to hurl stuff into space.

A brilliant solution to this problem would be a lightweight, super-strong cable stretching 100,000 kilometers from the Earth's surface out to a counterweight in space. Solar-powered elevators then would whisk people and freight into space for a fraction of the cost of rockets today.

The origin of this idea is nothing short of biblical. Writing in 1450 BC, Moses referred to an earlier civilization that had tried to build a tower to heaven out of bricks and mortar. Located in Babylon, it was called the Tower of Babel. In the Book of Genesis, there's the story of Jacob and his ladder. People have dreamt of a stairway to heaven ever since.

The first modern concept of a space elevator emerged from the fertile mind of Konstantin Tsiolkovsky. Inspired by the recently built Eiffel Tower, he imagined a structure 35,790 kilometers (21,475 miles) high that reaches the altitude of a geostationary orbit, where the orbital period equals the Earth's rotation period, so an object has a fixed location in the sky as seen from the ground. Release an object from the top of a tower this high and voilà, it's in orbit. But no material can handle the extreme compression created by such a structure, so Tsiolkovsky's idea languished. In 1959, another Russian scientist, Yuri Artsutanov, came up with a more feasible version. He suggested lowering a cable from a geostationary satellite while also extending a counterweight away from the Earth, to keep the forces in balance and the cable hovering over the same location on the Earth's surface. A space elevator would be kept in tension, with no compression or bending (Figure 33).

This is the physics. A space-elevator cable rotates with the Earth. So any object attached to it will feel an upward centrifugal force opposing gravity. Think of the outward force on an object tied to a string if you whirl it around your head; the object attached to the string acts as a counterweight, keeping the string straight and taut. The higher up the cable an object is, the weaker the Earth's gravity and the stronger the centrifugal force upward. The net gravity is less. At the geostationary orbital altitude, the centrifugal force acting upward perfectly balances gravity acting downward. So an object there experiences forces in perfect balance.

The space elevator, or tether, was reinvented a number of times in the 1960s and 1970s, and it entered the popular imagination with Arthur C. Clarke's 1979 novel, *The Fountains of Paradise*.[15] He realized that the cable should be tapered, thickest at the geostationary altitude and thinner at the ends, in order to equalize the amount of weight for a given cross-sectional area that the cable would have to bear. This is

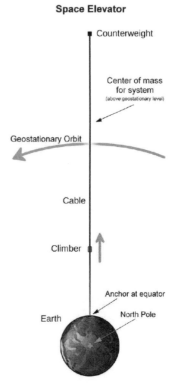

Figure 33. A space elevator has a cable fixed to the Earth's equator, extending into space. A counterweight at the open end ensures that the center of mass is above the level of a geostationary orbit. Centrifugal force keeps the cable taut.

because the cable has to be able to support, with tension, its own weight below any point. The location with the greatest tension is at the geostationary orbit level. Clarke also realized that as the lower sections of cable were built, the counterweight would have to extend to 144,000 kilometers, almost halfway to the Moon. Unfortunately, the engineers working on the problem realized that no known material could do the heavy lifting.

Materials that might be used for a space elevator are characterized by their tensile strength and density. A more useful figure of merit is their maximum length before they break under their own weight. We can start with natural organic materials. Jack's beanstalk couldn't handle the compression it would suffer going more than a few miles high. Natural fibers used in rope have good tensile strength, but their breaking distances are only five to seven kilometers. Steel cables used in bridges have breaking distances of 25 to 30 kilometers. These are well known from bridge building and other civil engineering projects. Spider silk has the same tensile strength as steel, even though it is a protein with just one-sixth the density, so the breaking length is 100 kilometers—impressive but still inadequate to get into low Earth orbit. Synthetic fibers like Kevlar and Zylon take us up to 300 or 400 kilometers, high enough to reach the ISS but not enough to send the counterweight much higher into space. The dreams of elevator operators looked like they couldn't be realized with conventional materials.[16]

Then nanotechnology burst on the scene in the 1990s. The ability to manipulate matter at the level of atoms or molecules opened up new technologies and a dizzying array of potential new applications. Some of the most exciting materials were made of pure carbon. Fullerenes are carbon molecules in the form of spheres, tubes, and other shapes. The name is a nod to the architect and designer Buckminster Fuller, since the first of the new molecules to be created was a tiny spherical cage made of sixty carbon atoms, resembling one of Fuller's geodesic domes. Soon after buckyballs were isolated, scientists learned how to create carbon nanotubes, interlinked carbon atoms rolled in a cylinder a millionth of a meter across. Carbon nanotubes are stable and they conduct heat and electricity extremely well.

But it was the mechanical properties that got space engineers excited. The tiny tubes are fifty times stronger than titanium; the theoretical limit is five times higher still. Carbon is the sixth lightest element in the periodic table, so there's almost no dead weight. Its strength and stability are unique for its low mass. The longest nanotubes are a few centimeters long, but if we can scale up the technology by a factor of a billion, we may be able to weave them into a carbon cable that reaches to the sky.[17]

We're certainly not there yet. Once, after a lecture, Arthur C. Clarke was asked when space elevators would become a reality. His response: "Probably about fifty years after everyone stops laughing."

Materials scientists disagree on whether carbon nanotubes are actually strong enough, and the technology needed to weave them into ribbons or ropes has never been demonstrated. If the hexagonal bonds become too strained, the structure can dramatically rupture, rather like a run in a woman's stocking. Such a long structure is susceptible to instabilities, whipping motions, and resonant vibrations. Moreover, the climbers or elevator cars create their own problems, inducing a wobble on the cable due to the Coriolis force (or Coriolis effect). The Coriolis force is familiar as the cause of weather systems rotating in opposite direction north and south of the equator. If you fly north or south from the equator, the ground moves at a slower speed under you, even though your speed hasn't changed. This leads to an apparent deflection as seen from the Earth's surface. A climber on a space-elevator cable would move more slowly on each successive part of the cable onto which it moved. This acts as a deflection or a sideways drag on the cable. The effect works in the opposite direction for a descent of the cable. In practice, this Coriolis force limits the speed at which a cable can be ascended.

Finally, there are risks from meteorites and from the 6,000 tons of space junk orbiting the Earth in its potential path, plus vulnerability of such a large target to a terrorist attack. One elevator isn't very efficient. There would have to be at least one for going up and one for going down. To avoid nasty oscillations, the speed might have to be kept to around 100 mph, making the journey take several weeks.

Space-elevator optimists are undeterred. A recently discovered carbon allotrope called carbyne is even stronger than the graphene that's the basis of current nanotubes.[18] It might be possible to "dope" the carbon to reduce the risk of ruptures. Space elevators got their most detailed design study ever in 2013 with a 350-page report from the International Academy of Astronautics.[19] The right material is still the wild card, but the report projects a space elevator carrying multiple 20-ton payloads by 2035. Getting international agreement for such a strategically important capability is a concern, as is protecting it from terrorist attacks.

The cost of anywhere from $10 to $50 billion is far cheaper than that of the International Space Station—for a tool that brings launch costs down to $100 per kilogram, twenty times lower than any rocket on the horizon. The new economic activity spawned by a space elevator could dwarf its cost.

A Space Boom

The history of space travel is fraught with unfulfilled promise. What is the basis for thinking that space will ever be more than a rarefied niche for those with deep pockets and nerves of steel? Hard-nosed economists have studied the issue and there's data to back up their claims.

The commercial space business is already a ubiquitous and mundane part of everyday life. Every time you use a phone to navigate to an unfamiliar location or watch a movie with the dish outside your house, you're using commercial space technology. It all started with a satellite the size of a beach ball, weighing no more than an average teenager. Telstar was launched in 1962 by NASA, but it was funded by AT&T, Bell Labs, the British Post Office, and the French Telecom Company. It's goal was to relay TV signals, phone calls, and fax images across the Atlantic. This was the birth of a global telecommunications industry.

GPS is the perfect example of the space industry in your pocket. We depend on our phones to give us the time and our location in any kind of weather and anywhere on the planet. This requires an unobstructed line of sight to four or more satellites in the Global Positioning System. GPS was developed in the 1960s by the US Department of Defense and was originally operated with twenty-four satellites. The military resisted giving civilians access to the system, fearing it would be used by smugglers, terrorists, or forces hostile to the United States. But in 1996 President Bill Clinton approved an upgrade to the current system of thirty-one satellites and committed to providing GPS tech-

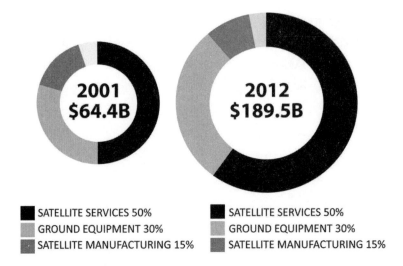

SATELLITE SERVICES 50%   SATELLITE SERVICES 50%
GROUND EQUIPMENT 30%   GROUND EQUIPMENT 30%
SATELLITE MANUFACTURING 15%   SATELLITE MANUFACTURING 15%

Figure 34. The bulk of commercial space revenue comes from launching satellites, where worldwide revenues nearly tripled between 2001 and 2012. Compared to this, space tourism is still a "minnow," but that will change if safe and reusable vehicles are developed.

nology worldwide and free of charge. A 2011 study showed that GPS technology sustains three million jobs in the United States and provides $100 billion each year in economic benefits.[20]

Launching satellites has become big business. In 2010, the FAA released a report on the economic impact of commercial space transportation: In the first decade of this century, it expanded from $64 billion to $190 billion, with a growth in jobs from half a million to a million. (For comparison, travel and tourism is three times larger and commercial aviation is six times larger.) It's now an international activity; fewer than half of the satellites launched for commercial use are built in the United States (Figure 34).[21]

The economic viability of space tourism is difficult to extrapolate—its capabilities aren't very impressive and its eventual size and long-term future are unclear. A few wealthy individuals have ponied up $20 million for a trip to the International Space Station, and it's the belief of space visionaries that as the price comes down, the demand will increase. But there are wild cards, such as the risk tolerance of people indulging in a recreation that could lead to a grisly end. The best market study done so far is by the Futron Corporation, an aerospace consulting firm with no skin in the space-tourism game. For orbital trips, they assume that the $20 million price tag will come down to $5 million after twenty years. Revenues would be $300 million at the end of that time frame. For suborbital trips, they assume a price of $100,000, declining to $50,000 after twenty years, when revenue would be a billion dollars a year.[22]

Surveys of the general public are consistent among industrialized countries; the lure of space knows no borders. If a brief trip to low Earth orbit cost only $10,000, about one million people would go, generating revenue of $10 billion per year. Interestingly, this is the same as the annual box-office revenues from movies in the United States. That's not bad, but those are still small potatoes when compared with $1.4 trillion spent on more mundane forms of tourism in 2013 (Figure 35).

The numbers get much bigger but also much more uncertain when we consider the mining of asteroids.[23] Terrestrial reserves are finite and there is growing exploitation of many elements that are critical for modern industry, such as antimony, cobalt, gallium, gold, indium, manganese, nickel, molybdenum, platinum, and tungsten. Many of these

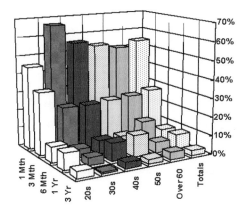

Figure 35. Market research from 1995 in the United States and Canada shows what proportion of their income people would spend on an orbital trip, broken down by age group. A third of respondents were not interested in a trip. The results can be used to model space tourism revenue.

strategic assets could be exhausted from accessible regions of the Earth in as little as fifty years. These ingredients were added to the Earth by a rain of asteroids 4.5 billion years ago, just after the crust cooled; if we want more of them, we'll have to go out and lasso more asteroids.

Space mining is in its infancy and it's very expensive. OSIRIS-REx is an asteroid study and sample return mission, funded by NASA and operated by the University of Arizona's Lunar and Planetary Laboratory. It's due to launch in 2016 and bring back as much as two kilograms of asteroid 101955 Bennu (named after a mythological Egyptian bird) in 2023. At $800 million, that's a bit expensive as a mining proposition, but the goal of this mission is to learn about the formation of the Solar System by retrieving pristine material from 4.5 billion years ago, when the Sun and all the planets formed.[24] As we've seen, NASA has a new plan to bring an asteroid the size of a large living room from deep space to an orbit around the Moon. That mission (with a price tag of nearly $3 billion) has been getting pushback from Congress, so it may not happen.

These concepts are small precursors to any viable mining operation. Despite the costs, however, the potential returns are eye-popping, according to plausible economic models.[25] In 1997, scientists estimated that a metallic asteroid a mile across contains $20 trillion of precious and industrial metals. Peter Diamandis, the X Prize guru who founded the extraterrestrial mining company Planetary Resources in 2012, has estimated that even a tiny, 100-foot-long asteroid holds as much as $50 billion worth of platinum alone. By 2020, he wants to build a fuel depot in space using water from asteroids to make liquid hydrogen and liquid oxygen for rocket fuel.

Experts remain skeptical. There's a huge difference between the market value of a space resource and the actual value after doing the hard work of mining the ore and bringing home the prize. As speculators have learned the hard way, cornering the market on a commodity can cause the floor on the price to collapse.

9

Our Next Home

Stepping Stone

Moon, Mars, and Beyond. The Moon is a hop, Mars is a skip, and the rest of the Solar System and the realm of the stars is a jump beyond. If we're to establish a permanent presence beyond the Earth, the best place to start is the Moon.

Almost forgotten in the more than forty years since we set foot on the Moon is the fact that NASA then had an aggressive plan for lunar exploration that would have culminated in a Moon base by 1980. Over the course of six Apollo landings, the time spent on the surface was extended from one to three days, the spacesuits were upgraded to allow moonwalks of up to seven hours, and the electric-powered rover was added to the mix. In 1968, as NASA prepared for its first piloted Apollo flight, it formed a working group to study the idea of a lunar station. After three exploratory missions to different landing sites, NASA would have sent six or more missions to a single site as preparation for a permanent base.[1] The working group began its report by declaring that a twelve-astronaut International Scientific Lunar Observatory should be a major goal for the agency (Figure 36).

Figure 36. An artist's concept of a lunar base. Most of the raw materials needed to build and supply a habitat could be mined or extracted from the lunar soil. No new technologies are required.

The working group's recommended option was to develop new hardware to form the nucleus of a future base. A new Lunar Payload Module with a descent stage but no ascent stage would carry 7,000 pounds to the surface. Its heaviest item of cargo would be a one-ton shelter capable of housing two people for two weeks. It would also deposit two snazzy personal jet packs for ranging over the surface and a rover or Moon buggy that could be driven by an astronaut or by flight controllers in Houston. The payload would also have included a solar furnace to test the extraction of useful ingredients from the soil, a one-foot telescope, a bioscience package, and various pieces of lab equipment. NASA's advisory group estimated that doing the groundwork for a lunar base would add a billion dollars to the projected cost of the Apollo program.

It was a great idea but it ran into the buzz saw of political reality.

NASA's budget peaked in 1965, during the white heat of development for the Moon landings, at $5.25 billion, or 5 percent of the federal budget. President Lyndon Johnson was a staunch NASA supporter, but the cost of the Vietnam War soared to $25 billion in 1967 and Congress was looking to cut costs. After the euphoria of Neil Armstrong's historic step, public interest waned and NASA's budget went into sharp reverse.

In 2009, the Center for Strategic and International Studies (CSIS) produced an estimate of the cost of a lunar base. They assume that a heavy-lift rocket will exist, and since at least three countries are likely to have such a capability, it's a fairly safe assumption. They project development costs of $35 billion, which is much cheaper than the $110 billion price of the ISS and, if spread over a decade, is no more than the costs of flying the Space Shuttle. Base operating costs are estimated at $7.4 billion per year. Half of the operating costs come from assuming that no local resources would be available, so four tons of supplies per person per year would have to be shipped from the Earth to the Moon. Basic requirements per day per astronaut (assuming water is efficiently recycled) would be 2.5 liters (or 2.5 kg) of water for drinking and adding to food, 0.8 kg of oxygen, and 1.8 kg of dried food. The other central requirement is energy in the form of solar power.[2]

Clearly, a lunar base would be more attainable if it could be as self-sufficient as possible. The Moon was always thought of as a sterile, arid, meteor-blasted rock, so there was much excitement when orbiters sent back evidence of water in the mid-1990s. In 2010, an Indian satellite found ice in the permanently shadowed regions of craters near the Moon's North Pole. This led to research showing that the Moon contains 600 million tons of ice in nearly pure sheets several meters thick.[3] The other key ingredient for a lunar base is oxygen to breathe. The lunar soil or regolith is 40 to 45 percent oxygen by mass; it's fairly simple chemistry to heat it to 2500 Kelvin using solar power and unlock it from minerals to generate 100 grams of breathable oxygen for every kilogram of soil. Water could also be split into oxygen and hydrogen, the main components of rocket fuel.

Even the material for a habitat could be created locally. Lunar soil is a unique blend of silica and iron-bearing minerals that can be fused into a glasslike solid using microwaves. Fairly simple technology can turn the dirt into hard ceramic bricks (Figure 37). The European Space Agency is developing a 3-D printer that can create wall blocks at three meters per hour, fast enough to build a whole habitat in a week.[4]

The cost of a Moon base shrinks dramatically if air, water, and build-

Figure 37. Cutaway view of a lunar base, based on an inflatable dome and a 3-D printing concept. After assembly, the inflated dome is covered with a layer of 3-D–printed lunar regolith to protect the occupants from radiation and meteorites.

ing materials can be generated locally. All of the technologies needed to do it have been demonstrated in the lab. There's no projection or wishful thinking of the kind invoked for space elevators, for example.

Location, location, location. It's as crucial when buying a home as when planning a Moon base. The best spots are high mountains on the rims of large craters near the poles. They would be close to abundant water ice but high enough to be peaks of eternal light, always illuminated by the Sun so with access to solar energy all the time. At low latitudes, colonists would have to contend with extreme temperature variations, plus the 354-hour-long lunar night. But they could make use of one of the tunnels formed long ago when basaltic lava flowed on the Moon. The lava tubes can be as wide as 300 meters. They maintain a stable temperature of –20°C as well as provide protection from cosmic rays and meteorites.[5]

A Moon base would be most valuable as a waypoint between Earth and Mars, and points far beyond. The Moon's low gravity and slow rotation mean that a space elevator could be built with materials already available. The honeycomb fiber called M5 is lighter and stronger than Kevlar; a ribbon 3 centimeters wide and 0.02 millimeter thick could support 2,000 kilograms on the lunar surface or 100 climbers with a mass of 600 kilos each, evenly spaced along the ribbon. We could build a lunar elevator right now.[6] The $35 billion development cost mentioned earlier does *not* assume a space elevator; it would reduce this cost by 20 to 30 percent.

Bill Stone, whom we met earlier with his plans to detect life on Europa, has formed the Shackleton Energy Company to prototype technology to separate lunar water into hydrogen and oxygen for rocket fuel. He knows that producing the fuel on the Moon will slash the cost of going elsewhere in the Solar System. To man his base, he won't be looking for typical NASA hires; he wants people with the spirit of wildcatters. To get back home, the first crew will need to produce the fuel for their journey.

With all this potential, it's safe to predict that the long hiatus in lunar exploration is over. In early 2014, China's Jade Rabbit rover arrived at the Bay of Rainbows in the northern part of the *Mare Imbrium,* or Sea of Rains. The private sector is showing interest. The Google Lunar XPrize was announced in 2007, with a $30 million grand prize to any team that lands a robot on the Moon and sends it 500 meters across the surface while sending back high-definition images and video. Eighteen teams are still in the running, and several might reach the Moon before the deadline of December 31, 2015.[7] India and Japan have plans for a lunar base by 2030; while Europe and the United States are currently dithering, they may collaborate on a base with a similar time frame.

Meanwhile, the space miners have their eyes on helium-3, an isotope of helium that's a key ingredient in a viable fusion power reactor. Helium-3 is extremely rare on Earth but more abundant in the lunar soil, where it's been generated over billions of years by the solar wind.[8] The idea is that fusion is the future of energy generation when coal, gas, and oil start running out, as they will around midcentury.[9] Fusion is a clean form of energy that leaves no radiative waste and has a high energy yield per kilogram of fuel. But fusion is extremely challenging. Europe and the United States each have been investing about $1 billion per year, with no energy-producing reaction sustained for more than a fraction of a second.

Most fusion research involves two isotopes of hydrogen: deuterium and tritium. The technical challenge is to contain the reaction, which takes place at a temperature of millions of degrees, far above the melting point of any known material. Helium-3 fusion requires even higher temperatures, but it has the advantage of generating most of the reaction energy in the form of charged particles, whose energy can readily be harnessed. The levelheaded view is that the Moon isn't going to rescue us from energy profligacy. Helium-3 presents three unproven realms of technology: harvesting it from the Moon, transporting it back to the Earth, and using it in a fusion reaction.

If we go back to the Moon, it will be for the more prosaic reason that it's the easiest and closest place to go to learn how to live off-Earth. American apathy might change if another country starts to colonize the Moon, and, as we've seen, the Chinese are thinking long term and making substantial investments in space projects. Early in 2014, the *People's Daily* quoted Zhang Yuhua, deputy general director of the Chang'e 3 lunar mission, as saying, "After the future establishment of the lunar base, we will conduct energy reconnaissance on the Moon, set up industrial and agriculture production bases, and make use of the vacuum environment to produce medicines."[10]

She closed by saying, "I believe that in 100 years, humans will actually be able to live on another planet."

The Lure of Mars

What happened to the Mars of our imagination?

The red planet used to be ominous and threatening. Babylonian astronomers took note of Mars, with its reddish color and its strange, episodic backward motion on the sky. They called it Nergal, god of the underworld and bringer of plague, epidemics, and disaster. The ancient Greeks associated Mars with Ares, one of the twelve Olympians and a son of Zeus and Hera. Ares was a violent and spiteful god who enjoyed combat; his sister Athene called him "a thing of rage, made of evil, a two-faced liar." The Greeks refused to honor him and no sacred places were built in his name. He was remembered on the battlefield, where his companions were Deimos and Phobos, the gods of terror and fear. His reputation softened slightly with the Romans, who made him the god of agriculture as well as war.

The Dutch astronomer Christiaan Huygens drew the first map of Mars in 1659; in a posthumous book titled *Cosmotheoros*, he speculated that the bright spots on Mars were evidence of water and ice. He also thought that intelligent life could exist there. A century later, William Herschel demonstrated that Mars had seasons like Earth's and argued that ice at the poles could support life. In the mid-nineteenth century, when large telescopes produced sharper images of Mars, some astronomers thought that the dark colorations might be due to vegetation and the striations or streaks on the planet might be artificial constructions.

Percival Lowell had no doubts at all. The Boston merchant and keen amateur astronomer used his fortune to build a new telescope at a pristine, dark site in northern Arizona. He was racing to complete the telescope in time for a particularly close approach of Mars in 1894. Lowell convinced himself that the features he saw on Mars were canals, where a dying civilization was trying to bring water from the poles to the equatorial regions. He wrote several sensational books to support his claims.

A few years later, H. G. Wells used aspects of Lowell's work in his science fiction novel *The War of the Worlds*. It was an instant classic.

Once again Mars was portrayed as malevolent: ". . . across the gulf of space, intellects vast and cool and unsympathetic, regarded our planet with envious eyes, and slowly and surely drew their plans against us."[11]

In the early twentieth century, the scientific and pop-culture views of Mars diverged. Edgar Rice Burroughs wrote *A Princess of Mars* as a serial in 1912 and as a book five years later. Civil War veteran John Carter mysteriously finds himself on Mars, a world populated with four-armed aliens, wild monsters, and scantily clad princesses. Carter uses weak gravity to exercise superhero powers and gets the girl in the end. Burroughs wrote ten more Mars stories, and his lurid fantasies inspired Arthur C. Clarke and Ray Bradbury to launch a grand tradition of Mars science fiction. In 1938, Orson Welles revisited *The War of the Worlds* with a radio show. His realistic live broadcast scared tens of thousands of people in the greater New York area; many raced from their homes at the prospect of a Martian invasion.

Meanwhile, Alfred Russel Wallace, codiscoverer of natural selection, had rebutted Percival Lowell, insisting that a freezing Mars could never support liquid water. This argument got stronger with remote sensing in the middle of the twentieth century. Mars fever finally cooled in 1965, when Mariner 4 swooped within 10,000 kilometers of the planet's surface and saw an arid, crater-pocked terrain with no signs of life. The twin Viking landers cemented this picture in 1976.

Since then, however, the pendulum has swung back toward the middle. Mars can't host liquid water on the surface; a cup of water on the surface would evaporate in seconds because the typical temperature is −60°C. It would evaporate rather than freeze because the atmosphere is so thin, very close to vacuum. The topsoil can't host life because the thin atmosphere allows it to be blasted by micrometeorites and sterilized by ultraviolet radiation and cosmic rays. But orbiters have provided abundant evidence of erosion and river deltas and shallow seas in the past. Craters overlying these features suggest that three billion years ago Mars was warmer and wetter and had a thicker atmosphere. A series of intrepid rovers—the Tonka Toy–size Sojourner in 1997, the go-kart–size twins Spirit and Opportunity starting in 1993, and the SUV-size Curiosity more recently—have painted this picture in more detail. These rovers have collected rock samples that can only have formed in the presence of water.[12]

Astronauts on Mars wouldn't find water on the surface but they'd be able to gather all they needed by a careful choice of landing site. Spectroscopy from orbiting spacecraft reveals extensive ice at high latitudes, hidden by a veneer of dust and rock. If melted, and if there was a protective atmosphere, it would cover the planet with a puddle deep enough to get your ankles wet. There's also evidence of gullies, with channels carved by water that erupts episodically from aquifers under the surface, where water can be kept liquid by pressure from above and radioactive heating from below (Figure 38). Scientists debate how deep colonists would have to drill to hit water, but it might be as little as 10 or 20 meters.[13]

While it's not the Mars of anthropomorphized aliens or comic-book superheroes, it's a planet that could host microbial life right now, as

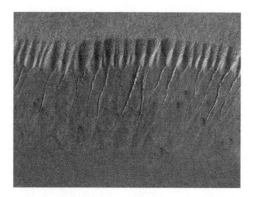

Figure 38. The sharp V-shaped gullies on this escarpment on Mars are strong evidence of liquid water runoff. The water is likely to have seeped out of the escarpment about a third of the way down from the top, where it can be kept liquid by pressure in aquifers.

well as in the past. The red planet is far more habitable than the Moon because of the atmosphere, the higher temperature, and the amount of subsurface water. It beckons us to visit, and perhaps stay.

Establishing a Colony

Robert Zubrin never lost the faith. With a PhD in nuclear engineering and more than 200 technical papers to his credit, Zubrin has been a staunch advocate of human exploration of Mars for thirty years. He holds patents for hybrid rocket planes, synthetic-fuel manufacturing, magnetic sails, saltwater nuclear reactors, and three-person chess, but his true passion is Mars. He thinks we can lower the cost and complexity of a Mars mission by "living off the land," or utilizing as many resources as possible from the air and soil. His ideas were strong enough to be adopted by NASA as their "design reference mission," but he became frustrated at NASA's glacial progress and anemic government support, so he founded the advocacy group Mars Society in 1998. He's written a series of books that make the case for going to Mars.[14]

Asked about saving costs with a one-way journey, Zubrin has said: "Life is a one-way trip, and one way to spend it is by going to Mars and starting a new branch of human civilization there."[15]

Mars is a challenging goal for human exploration. The problem isn't energy. The energy cost of going to Mars is less than 10 percent more than the energy cost of going to the Moon. The problem is the distance. An energy-efficient trajectory involves a travel time of nine months each way. The trip can be shortened to six or seven months at the expense of extra energy—a far cry from the week it takes to get to the Moon. The cost of transporting two years' worth of supplies for even a small crew is daunting. Wernher von Braun was the first to make a technical study of a Mars mission in the 1950s, but it was hopelessly grandiose, using a thousand Saturn V rockets to build a fleet of ten spacecraft in Earth orbit to then carry seventy astronauts to Mars. He pitched a scaled-down concept to President Richard Nixon, but it was passed over in favor of the Space Shuttle. Former NASA administrator Thomas Paine tried next. Perhaps he'd watched too much *Star Trek*, but he aimed to conquer and industrialize the Moon with nuclear space tugs, launch a fleet of space stations into Earth orbit, and send several dozen space-ships a year to Mars to build a space station and support the colony. The Reagan administration was happy to shelve his report.

In 2014, the National Research Council revisited human flight, as directed by Congress. Its sweeping 286-page report concluded bluntly that NASA had an unsustainable and unsafe strategy that will prevent the United States from achieving a human landing on Mars any time in the foreseeable future.[16] With current budgets, they suggest that it can't happen before midcentury. Along the way, the report addresses the philosophical question of why we should send people into space at all, concluding that purely practical and economic benefits don't justify the cost, but the aspirational aspect of the endeavor might make it worthwhile.

There must be good reasons and a strong will, because Mars is hard.

One risk is radiation. Earth dwellers are sheltered from high-energy cosmic rays and solar flares by our atmosphere and magnetic field. When the Curiosity rover headed to Mars, scientists switched on a radiation detector and found that the radiation environment in deep space is far more intense than it is on Earth. An astronaut on a two-year trip to Mars would get a 200 times greater radiation dose than an Earth dweller over that same period (Figure 39). However, to keep it in perspective, the adventure only increases the lifetime risk of cancer from 21 percent to 24 percent. The risk of spacecraft malfunction is likely to be much higher.

Another risk is weightlessness. We've talked about the substantial physiological changes resulting from a microgravity environment. Russian cosmonaut Valeri Polyakov spent 438 days aboard Mir, making a dizzying 7,000 orbits of the Earth, in part to see whether humans could handle a trip to Mars. The Russians reported that he suffered no long-term ill effects from his fourteen months in space. Robert Zubrin notes that the used upper stage of a Mars launch vehicle could be employed as a counterweight. With a mile-long tether and a spin rate of 2 rpm, Earth gravity would be simulated. With a spin rate of 1 rpm, it would

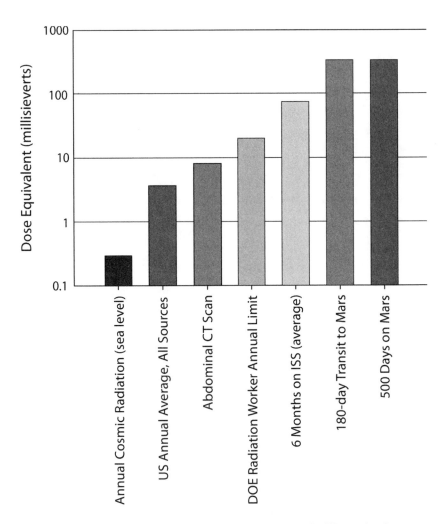

Figure 39. Comparison of the exposure to high-energy cosmic rays in different situations; note the logarithmic scale. Most of the exposure comes from the one-year transit time, equivalent to a hundred years of normal terrestrial exposure.

Figure 40. Design of a proposed Mars base from a concept called Mars Direct, developed by NASA engineers Robert Zubrin and David Baker in 1990 using only proven technologies. The Manned Habitat Unit is docked alongside a similar pre-placed unit that was sent ahead of the Earth Return Vehicle.

be Mars gravity and the astronauts could get acclimated to the new situation before landing (Figure 40).

A third risk is being cooped up. Mars travelers would have to spend a year and a half in a cabin the size of a school bus, and as much as a year at their destination in a space no bigger than a large motor home. The Mars500 mission locked an international crew of six volunteers in a mock spaceship theoretically bound for Mars—when in fact they were sitting in Moscow for a year and a half. The crew "returned to Earth" in 2011. Most of them experienced severely disrupted sleep patterns and all of them reduced their activity levels in the confined space, something researchers call a behavioral torpor.[17] The experiment made clear how important it will be to simulate Earth life rhythms in the spaceship or on Mars, and how important it will be to stay physically active.

It's hard to judge the psychological impacts of such a trip. People who winter in Antarctica experience a diluted version of the problem, but travelers to Mars will be the most isolated humans who ever lived.

They'll have real-time interactions with a small number of companions and delayed communications with friends and loved ones who are tens of millions of miles away. They'll be in a confined space with no option to simply go out for a walk, and they'll be monitored continuously by anxious ground crews and scientists on Earth. If anyone spins out of control, there's no real-time access to mental health services such as counseling or psychotherapy.

The visionaries are undeterred. Apollo astronaut Buzz Aldrin put it like this: "Going to Mars means staying on Mars—a mission by which we are building up a confidence level to become a two-planet species. At Mars, we've been given a wonderful set of moons which can act as offshore worlds from which crews can robotically preposition hardware and establish radiation shielding on the Martian surface to begin sustaining increasing numbers of people."[18]

Two new ventures are trying to put Mars within reach without using any government resources. Inspiration Mars is the brainchild of Dennis Tito, the engineer-turned-tycoon who was the world's first space tourist in 2001. Tito plans to keep costs down by not landing. His billion-dollar flyby plans to use an upgraded version of the SpaceX Dragon capsule. With a cleverly designed trajectory, he can get there with a single burn of the engine. The return is challenging, however. The capsule will slam into the Earth's atmosphere at 32,000 mph, requiring new materials for a heat shield. The project is aiming for a launch in 2021.[19]

Mars One is run by Dutch entrepreneur Bas Lansdorp, who also plans to use a SpaceX capsule. He will keep costs down by leaving his four passengers on Mars. If they survive the trip, they will build a habitat from their spacecraft and adjacent inflated areas covered by Martian regolith. They'll create water, oxygen, and some food locally, augmented by regular supply missions, and every two years they will be joined by four more refugees from Earth. Gradually, they will build a colony. Lansdorp estimates his costs to be $6 billion for the first trip and $4 billion for each crew that follows. Space experts judge the plan to be very ambitious; some judge it to be impossible. Everyone agrees it's audacious.[20]

Would-be Martians are in a race against time. The red planet has its next close approach to the Earth in 2018, and it won't get as close again until 2035. Inspiration Mars and Mars One have both had to slip past the most favorable 2018 launch date. Mars One accepted more than 200,000 applications online for the chance to live and die on Mars. In 2014, that number was culled to 1,058 from 107 countries, and then to 705. Those who remain will endure rigorous physical and psychological testing to generate a final group of twenty-four. Lansdorp plans to finance his venture by turning it into a reality TV epic—think *Survivor* meets *The Truman Show* meets *The Martian Chronicles*.[21]

Greening the Red Planet

Let's ignore for a moment the evil twin. Venus is closest to the Earth in size and mass, and it has the same inventory of carbon dioxide. But on our planet most of the carbon dioxide is built into rocks and dissolved in the oceans, making them mildly acidic and leaving a moderately thick atmosphere to smooth out daily and seasonal temperature variations.

On Venus, only 30 percent closer to the Sun, the carbon dioxide built up in the atmosphere, triggering a runaway greenhouse effect and raising the surface temperature to a level where lead melts. Whoever named Venus after the goddess of love had a sad history of relationships.

Mars is the misbegotten sibling, the runt. It's half the size of Earth with a third of the gravity. The next nearest Earth-like planet is tens of trillions of miles away, and unreachable with any current technology. The siblings went on divergent paths. One rusted and turned red, the other got the spark of life and turned green. Mars suffocated and dried out as its water and air leached into space, and it became scoured by dust storms and cosmic rays. Yet it's at the edge of habitability, not much more inhospitable than a volcanic vent or a plateau in the high Andes. Mars has sunlight and reservoirs of water, carbon, nitrogen, and oxygen. One planet lived and the other died.

Perhaps we can make it live again?

One of the most audacious ideas in science is planetary engineering. Planets don't stay the same. Geological evolution, combined with the aging of their parent star, can render a wasteland habitable and an Eden uninhabitable. This evolution occurs on geological timescales of hundreds of millions or billions of years.

Here's how the Earth has changed. It formed 4.5 billion years ago and minerals show that there was liquid water within 100 million years, so conceivably life started then. If it did, it must have survived the "Late Heavy Bombardment" 3.9 billion years ago, when unstable orbits in the Solar System led to a surge of meteor impacts. Life around that time was limited to prokaryotes, or cells without nuclei, and there was no oxygen in the atmosphere. Around three billion years ago, bacteria evolved that produced oxygen as a waste product, which is poisonous to other kinds of bacteria. The oxygen content of the atmosphere rose 1.9 billion years ago and facilitated the evolution of eukaryotes, or cells with nuclei. Life diversified as it became multicellular and began to reproduce sexually. Dramatic episodes of glaciation almost obliterated life 2.7 billion and 700 million years ago. In the last 10 percent of the chronology, life finally became big enough to see without a microscope, plants and animals evolved, they moved onto the land, and a crescendo of evolution led to mammals, primates, and finally us. Dramatic change is normal for a biological world.[22]

More recently, we have been inadvertently altering our own planet through industrial growth and the use of fossil fuel. "Terraforming" is the process by which we might potentially alter a different planet to make it more Earth-like or habitable by terrestrial life forms.

The first step would be to raise the temperature on Mars just enough to release frozen carbon dioxide from the polar regions, triggering a runaway greenhouse effect. The positive feedback of this effect favors terraforming. While the carbon-dioxide atmosphere of Mars has only 1 percent of the pressure of the Earth's atmosphere at sea level, there is enough carbon dioxide frozen in the soil to raise the pressure to 30 percent of the Earth's. Robert Zubrin and Chris McKay have outlined several ways to accomplish this. Chris McKay is a NASA astrobiologist who believes we have an obligation to seed life on planets that might be habitable. One strategy is to fabricate a 100-kilometer mirror to direct extra sunlight toward the poles. Even if made from aluminized Mylar, such a mirror would weigh 200,000 tons. Being too heavy to launch from Earth, this would have to be constructed from materials refined on Mars.

Another method is to produce efficient heat-trapping gases on Mars, using industrial-scale facilities. There's a rich irony in using methods to make Mars habitable that are in danger of rendering the Earth uninhabitable. These two methods would each use as much energy as a city like Denver or Seattle, and they would need hundreds of workers to implement. A clever, less costly idea is to redirect small asteroids to impact the surface of Mars. Carbon dioxide would be liberated by heat energy from the impact, and asteroids can deliver ammonia (a very efficient greenhouse gas) and dust, which will cause Mars to absorb more sunlight.[23]

The next step is to activate a hydrosphere: raise the temperature by an additional amount sufficient to allow liquid water on the surface. Although still inhospitable, these conditions would allow extremophile microbes such as lichen, algae, and bacteria to be established. Their role is to prepare the regolith for photosynthetic organisms. Microbes used for this will be engineered to be optimally suited for their job. If the heating is done with asteroid impacts, these first two steps might take two to three hundred years.

The last step is to add oxygen to the atmosphere. Since oxygen is flammable, care would have to be taken to also add a buffer gas like nitrogen. Brute force would have to be used to import or create the initial oxygen needed for primitive plants, but when more advanced plants can propagate, they become the engine for oxygen production. It would take 500 to 1,000 years to make an atmosphere suitable for animals or humans.

Terraforming may be possible and it's exciting at a technical level, but to see life breathed into the idea, we can turn to fiction. Kim Stanley Robinson wrote a science fiction trilogy in the mid-1990s about an overpopulated and dying Earth and the "First Hundred," a pioneering group of Mars colonists. The books capture the ethical issues we'll face if we go there, telling of the tensions between the Reds who prefer to leave Mars in its pristine state and the Greens who want to turn the planet into a second Earth.[24]

The storytelling is very entertaining, but the physical descriptions are beyond evocative; they're mesmerizing. Who wouldn't want to visit Mars after reading this excerpt from *Red Mars*, the first book in the trilogy: "The sun touched the horizon, and the dune crests faded to shadow. The little button sun sank under the black line to the west. Now the sky was a maroon dome, the high clouds the pink of moss campion. Stars were popping out everywhere, and the maroon sky shifted to a vivid dark violet, an electric color that was picked up by the dune crests, so that it seemed crescents of liquid twilight lay across the black plain."

10

Remote Sensing

Extending Our Senses

What if we could have the experience of space travel without actually making the journey?

The cost and difficulty of protecting fragile humans and sending them vast distances through space suggest that we should find a different way to explore space. To see an alternative, look at the evolution of video games.

Pac-Man was the most famous arcade video game of all time. Released in 1980, the game had the player steer a small colored icon through a maze eating dots. Pac-Man's popularity eclipsed that of space shooter games like Space Invaders and Asteroids, and it's been estimated that by the end of the twentieth century, ten billion quarters had been dropped in Pac-Man slots. In 2000, a new computer game came out in which the player could create virtual people, houses, and towns and watch their cartoon characters live their virtual lives. The Sims sold more than 150 million copies worldwide. If we think of how far video games came in twenty years, from the primitive graphics of Pac-Man to the cartoonish but quasi-realistic 3-D graphics of The Sims, imagine what another twenty years will bring. A hint of that came in 2014 with the release of the Oculus Rift, a gaming helmet that immerses a player in 3-D virtual reality.[1] The best sense of the experience is the dramatic opening sequence of the 3-D movie *Gravity*.

The future of Solar System exploration may lie in telepresence, a set of technologies that allow a person to feel that he or she is in a remote location. Videoconferencing is one familiar and simple form of this technology. The market for projecting images and sound to connect meeting participants from around the world is growing 20 percent a year and is worth nearly $5 billion. Skype video calls now account for a third of all international calls, a staggering 200 billion minutes a year. Other examples include using robots with sonar to explore the ocean floor or robots with infrared sensors to explore caves. The robot provides the "eyes and ears" for an operator who doesn't have to leave the comfort of an office or home.

When we "look" at Mars through the camera eye of the Curiosity rover or "sniff" the atmosphere with its spectrometer, we are using a form of telepresence. NASA has used red–green stereoscopic imaging on all of its recent rovers, but it missed a big chance to grab the public eye when it failed to build a 3-D high-definition video camera into the Curiosity rover in time for launch. Film director James Cameron had pitched the camera to give Earthlings a "you are there" immediacy as the rover trundled around the red planet.[2]

A lot has changed since the last Apollo astronaut walked on the Moon. At the time of the first Moon landing, real-time, complex decision making had to be carried out by people. Now, robots and machines have impressive capabilities, so they can be remotely controlled by scientists at great distances.

Planetary scientists have used remote sensing for some forty years. The twin Viking landers were designed to analyze samples of Mars soil for traces of microbial life. No cameras were included in the specifications, but Carl Sagan argued that images from the surface would engage the public. Besides, he noted mischievously, what if there are Martian polar bears and we miss them because we don't take pictures? So cameras were added, and their images of stark desert vistas were immediately compelling to the public. Probes to the outer Solar System since then have "watched" the volcanoes on Io, "listened" to magnetic storms on Jupiter, "sniffed" the atmosphere of Titan, and "tasted" the icy geysers on Enceladus.

Telepresence implies something more than remote sensing; it's a technology that allows someone to feel as if he's in a remote location. The word was coined in 1980 by US linguist and cognitive scientist Marvin Minsky. He was inspired by a short story by science fiction author Robert Heinlein. The concept was further developed by Fred Saberhagen in *Brother Assassin*, from the Berserker series:

> . . . it seemed to all his senses that he had been transported from the master down into the body of the slave-unit standing beneath it on the floor. As the control of its movements passed over to him, the slave started gradually to lean to one side, and he moved its foot to maintain balance as naturally as he moved his own. Tilting back his head, he could look up through the slave's eyes to see the master-unit, with himself inside, maintaining the same attitude on its complex suspension.[3]

This level of control and verisimilitude is far off in space exploration, but we're approaching it with the virtual reality of video games. The difference between gaming and science applications is that a video game tries to digitally re-create a real-world experience while science uses technology to digitally represent and transmit the real world.

Remote control of robots—often called telerobotics—is infiltrating life in surprising ways. Robots are used nowadays to defuse bombs, extract minerals from hazardous mines, and explore the deep sea floor. They also act as aerial drones and doctor's assistants. They're even beginning to be seen in the boardroom and the workplace. Many commercial robots look like vacuum cleaners with a screen on top and are no more than ventriloquist's dummies; after the comical first impression, it's disconcerting to realize that there's a real person at the other end of the device. A striking recent example was a talk by Edward Snowden at the TED2014 conference.[4] The controversial NSA whistleblower was in hiding somewhere in Russia, but he was represented on stage by a screen attached to two long legs that ended in a motorized cart. Snowden communicated with the moderator and turned toward the audience to answer questions; he could see and hear everything that was going on.

At a 2012 symposium on "Space Exploration via Telepresence," held at NASA's Goddard Space Flight Center, scientists rubbed shoulders with roboticists and technology entrepreneurs. A major topic was latency—the time it takes a robot to respond to commands and communicate results back to the operator. Latency is governed by the speed of light. In terrestrial applications, latency is essentially zero, but on the Moon it's a couple of seconds, on Mars it ranges from ten to forty minutes, and to the outer Solar System it's up to ten hours. This makes real-time communications impossible.

Astronauts on the International Space Station have tested the remote control of a mobile robot named Justin, which was developed by the German Aerospace Center.[5] The robot has four-fingered hands, and astronauts control it with a sense of "touch." This is done with haptic technology that uses forces and vibrations to re-create the feeling of touch.[6] To avoid latency, and to avoid the costs of going in and out of gravity pits, future explorers may control ground operations from Moon orbit or Mars orbit. NASA is testing a "blue collar" robotic miner that digs, fills and empties buckets, and can right itself if it falls. It would be part of the advance expedition to Mars to mine and build with local materials in preparation for the later arrival of astronauts. Meanwhile, the European Space Agency is developing a robotic exoskeleton so that

Figure 41. Robonaut is a robotic humanoid development project conducted by NASA's Johnson Space Center. This version, R2, was first used on the International Space Station in 2011. The torso can be positioned to help the crew with engineering tasks and extra-vehicular activities (EVAs).

astronauts can control a remote robot as if it were an extension of their body, and they've already tested a robot that can carry out simple tasks aboard the International Space Station (Figure 41).

The frontier of telepresence is its merger with artificial intelligence, a development foreseen by computer science pioneer Marvin Minsky in 1980.[7] A robot doesn't need to be just a remote extension of a human; it can process information and make its own decisions. This will be exciting, but it will raise fascinating moral and ethical questions, especially if these semiautonomous robots come into contact with each other.

Here Come the Bots

Richard Feynman was an iconic physicist who won a Nobel Prize for his work in quantum theory. His delight in understanding how nature worked was infectious. In 1959, he wrote an influential essay titled

"There's Plenty of Room at the Bottom," in which he argued that miniaturization of computers still had a long way to go. He talked about the limits of making machines and computers and realized that there might one day be technologies that could manipulate matter on the scale of individual molecules and atoms.[8]

That day has finally arrived.

Nanotechnology involves scales of a billionth of a meter or smaller. It's disconcerting to think that our world might one day be run by robots too small to see, but the benefits could be enormous. We're familiar with swallowing a pill to treat an illness or a disease. But what about swallowing a pill-size robot that can monitor our vital functions from inside and warn of impending problems? Or a pill that could release a thousand tiny molecular machines to combat microbes or regenerate bones or blood vessels? Feynman anticipated a time when we could "swallow the doctor."[9]

Some people think nanotechnology is unnatural, but the research is often inspired by biology. A beautiful example is flagella, which propel bacteria in a liquid medium. They're molecular motors, complete with propellers, universal joints, rotors, gears, and bushings.[10] Cancer therapy is high on the list of medical applications, since the current treatments rely on drugs and radiation, which are blunt and often toxic tools. Nanobots would be able to move directly to a cancer site, distinguish between malignant and normal cells, and do treatment without side effects or damage to the immune system. The potential of using nanobots for drug delivery and regenerative medicine has galvanized medical researchers. Federal grants for applying nanotechnology to medicine now exceed $2 billion (Figure 42). Similar capabilities will be focused on the environment, where nanosponges can be used to clean up oil spills and neutralize toxic chemicals. These nanosponges could also increase the efficiency of oil extraction, avoiding adverse effects of fracking.

The US military is investing heavily in nanotechnology. Many of the programs are classified, but they include miniaturization of drones to

Figure 42. Nanobots or micromachines will be used increasingly in medicine to deliver drugs, repair damaged tissue, and fight diseases such as cancer. These same technologies can be used for exploration of planets and their moons.

the size of insects to conduct surveillance, the use of "smart motes" the size of grains of sand to monitor battlefields for toxic gases, and development of armor and protection for soldiers that can alter its structure at the molecular level.[11]

So what does nanotechnology imply for space exploration? The big difference between us and our machine proxies is that robots can be shrunk and we can't. Nanobots are too small for batteries or normal solar cells, so tiny amounts of radioactive materials would be used to power them. Conversely, nanotechnology is leading to new, efficient designs of solar cells, so nanobots could be used to assemble solar panels at a remote location.

The Curiosity rover is a wonderful machine, but this SUV-size Mars vehicle will seem like a dinosaur when nanobots arrive on the scene. The first wave would be dropped from an orbiting spacecraft—in weak Mars gravity, they would ride the wind like a dust storm of smart motes. Each nanobot would have a processor, an antenna for communicating with other nanobots and the orbiter, and various sensors. The skin of each nanobot would be a shape-shifting polymer to optimize drifting on air currents or being blown along the surface. All these capabilities have been prototyped at millimeter scales; there are no fundamental obstacles to shrinking them further.[12]

For more complex missions, robots would move independently under their own power. NASA researchers have developed the concept of ANTS, or Autonomous Nanotechnological Swarm—robots a millimeter across made of a tetrahedron of carbon-nanotube struts connected by joints. Each robot would move by shortening or lengthening its struts, altering its center of gravity to tumble in the right direction. Imagine thousands of these tiny rovers, linked by a neural net, fanning across the surface and carrying out geological tests to look for signs of life. NASA recently auctioned off the patent for the robots, hoping to spur innovation.[13]

Nanobots could also convert carbon dioxide into oxygen; if they were self-replicating, they could greatly accelerate the process of terraforming Mars. And not just Mars. The same NASA research group thinks nanobots made of carbon nanotubes could survive the 900-degree temperature on the surface of Venus. Following on from the billion-dollar, multi-ton Cassini mission, a spacecraft the size of a shoebox could seed remote-sensing nanobots on Titan and Enceladus. With asteroids, they could do an inventory of precious metals and rare earths as the prelude to full-scale mining.

Nanobots can also help humans be safer when they explore space. Constantinos Mavroidis, a professor of engineering at Northeastern University, formed a team to map out concepts that are attainable within forty years. One of their ideas was a strong but lightweight spacesuit that could repair itself with protein-dispensing nanounits built into the layers of fabric. The same suit could also monitor the vital signs of astronauts and carry emergency drugs. When there's room at the bottom, the sky's the limit.[14]

Solar Sailing

The cost and difficulty of space travel is rooted in dependence on the way rockets work. Sunlight is used to generate electricity and power on Earth and in space. What if it could be used for propulsion as well?

It can. Johannes Kepler noticed that comet tails point away from the Sun, and in a 1610 letter to Galileo, he suggested: "Provide ships or sails adapt to the heavenly breeze, and there will be some who brave even that void." Jules Verne was the first to outline the concept of the solar sail in 1865, seizing on James Clerk Maxwell's theory that light has momentum as well as energy so it can exert pressure on objects. In *From the Earth to the Moon*, Verne wrote: "Light or electricity will probably be the mechanical agent [by which we shall] travel to the moon, the planets, and the stars. . . . "[15]

The physics is simple.[16] When photons hit any reflective surface, their momentum changes as they reverse their velocity, imparting a very small unit of force to the object they hit. When the shiny surface of a sail faces the Sun, the sail is pushed away from the Sun. Solar sails can do anything nautical sails can do, including tacking. Changing the angle of the sail relative to the Sun affects the direction in which the sail is propelled. The sail can direct a spacecraft toward the Sun by using photon pressure to slow the spacecraft and lower its orbit. A solar sail can even act like an antigravity machine, using solar pressure to balance the Sun's gravity and hover anywhere in space.

Light pushes very gently on a solar sail in Earth orbit, so acceleration is feeble when compared with the kick imparted by a rocket. But a rocket will run out of fuel while the Sun keeps shining. This steady stream of photons creates acceleration that adds up to a substantial velocity boost. A solar sail should be big and light, to catch as much sunlight as possible and thus impart maximum possible velocity to the spacecraft. Cosmos 1 was intended to be a prototype for interplanetary travel using a solar sail. It was funded by the Planetary Society and Cosmos Studios, the film studio founded by Ann Druyan, Carl Sagan's widow. Sagan had died in 1996 and she wanted the 600-square-meter Mylar sail to be his Taj Mahal, a monument to a man who had argued for us to weigh anchor and set sail for the stars.

Sunlight would exert a tiny acceleration of half a millimeter per second per second on the huge sail. After one day, the speed would be just 100 mph, but after 100 days, its speed would be 10,000 mph. After 1,000 days, it would reach a blistering 100,000 mph. Unfortunately, the Volna rocket that was launched from a Russian submarine in June 2005 failed, and Cosmos 1 went to the bottom of the Barents Sea.[17]

Solar-sail development has continued, but the ambitions and the size of the sails have been scaled back. A team from NASA built NanoSail-D, based on the CubeSat specifications. CubeSat is a miniaturized satellite designed to spur space research by using standard components and off-the-shelf electronics. A CubeSat is a bit bigger than a Rubik's Cube—10 centimeters on a side and weighing less than 1.3 kilograms. Most CubeSat launches have come from academia, but companies such as Boeing have built CubeSats, and amateur satellite builders have gotten their projects off the ground using crowdfunding campaigns on websites such as Kickstarter.

NASA's NanoSail-D was designed to use three CubeSats to deploy triangular sails totaling 10 square meters. Unfortunately, it too was scuppered by the launch vehicle when its Falcon rocket malfunctioned in 2008. But NASA persisted, and a twin was successfully launched in 2011 (Figure 43). NanoSail-D was never intended to be more than a test of solar-sail deployment, and it burned up after 240 days in low Earth orbit. The previous year, the Japan Aerospace Exploration Agency (JAXA) had sent IKAROS toward Venus. It was the first spacecraft to be fully powered by a solar sail. NASA had plans to launch a 1,200 square-meter sail called Sunjammer but

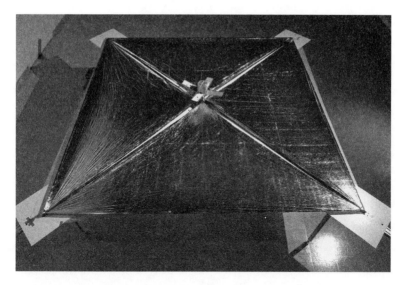

Figure 43. This solar sail developed by NASA has a light-catching surface of 100 square feet, and the sail and the spacecraft together weighed less than 10 pounds when it was deployed in 2011. The NanoSail-D structure is made of aluminum and plastic.

cancelled the project in late 2014.[18] The name comes from an Arthur C. Clarke short story.

CubeSats are at the epicenter of the business plans of commercial space companies. In the next five years, more than a thousand nanosats will be launched, some larger than a CubeSat and some smaller.[19] In early 2014, a batch of twenty-eight CubeSats was released by a satellite "shooter" aboard the International Space Station to take pictures of the Earth. In 2013, the first PhoneSat went into orbit. This used sensing plug-ins for a Google smartphone to measure magnetic fields, pressure, and more. If the launch cost drops below $1,000 per kilo, anyone will be able to get into the game. Nanosats will be the first choice for remote sensing on moons and planets in the Solar System.

Like space elevators, solar sails are still in their infancy. The challenge is to launch a gossamer film that's reflective, the size of a football field, and a hundred times thinner than a sheet of paper. The sail has to be launched in a compact configuration and then unfurled in space and held rigid by a frame or inflatable boom. Solar sails are slow and steady in their acceleration, but they face diminishing returns as they reach the outer Solar System, because the amount of "push" from solar photons goes down by the square of the distance from the Sun. The sail doesn't slow down, but its speed does increase more and more slowly.

As a result, some people are exploring radical ideas. An electric solar sail doesn't look like a sail; rigid conducting wires extend radially from the spacecraft and a current keeps them charged up to 20,000 volts. Their electric field makes them look 50 meters thick to the solar-wind ions, and that interaction drives the spacecraft. Magnetic solar sails also use the solar wind, but they deflect the ions with a magnetic field made by running current through wire loops. Magnetic sails can use planets and the Sun for thrust, by pushing against their magnetic fields.[20] To venture beyond our Solar System "harbor," we'll want to build up as much speed as possible before coasting across the vast interstellar sea.

Finding Alien Technology

We've had the ability to leave the Earth for two generations, a tiny fraction of the thousands of generations since we made our epic journey out of Africa. We share intelligence, and perhaps sentience, with species such as chimps, dolphins, orcas, and elephants. But we're the only species that has bent the material world to our will and fashioned computers and skyscrapers and rockets. Are we the only creatures who have developed the technology to venture beyond their home planet?

The best way to answer this question is to use the fastest thing there is: electromagnetic waves.

Remote sensing has let us diagnose distant planets and look for signs of microbial life. It can also let us vault over uncertainties in biological evolution and look for the hallmarks of intelligence and technology. For decades, science fiction writers have woven stories of aliens who are biologically bizarre or who have eclipsed us with their technology. Scientists play this game too, and it's known by the acronym of SETI: Search for Extraterrestrial Intelligence.

In an influential paper published in *Nature* in 1959, "Searching for Interstellar Communications," Giuseppe Cocconi and Philip Morrison argued that a search was warranted even though there was no evidence for life any place other than Earth. They wrote: "The reader may seek to consign these speculations wholly to the domain of science-fiction. We submit, rather, that the foregoing line of argument demonstrates that the presence of interstellar signals is entirely consistent with all we now know, and that if signals are present the means of detecting them is now at hand."[21]

They argued that we target nearby Sun-like stars and look for narrow-bandwidth microwave signals. Radio waves aren't naturally produced by stars, so radio waves appearing from a star could only come from an artificial source next to the star. Radio telescopes and powerful radio transmitters had been developed a decade earlier. Visible light isn't the best choice for signals because thick planet atmospheres are opaque and billions of stars in the galaxy present a confusing noise source. The radio regime is much quieter because stars don't emit radio waves. Moreover, there's a particularly quiet zone in the cosmic environment between 1 GHz and 10 GHz where water vapor doesn't absorb radio waves, so they travel freely across large distances in the galaxy. It also happens to be the region where hydrogen has a fundamental spectral transition that would be noteworthy to any alien civilization that knew about physics.

Cocconi and Morrison urged astronomers to search this region, and to doubters they noted: "The probability of success is difficult to estimate, but if we never search, the chance of success is zero."

Soon afterward, a young researcher named Frank Drake pointed the 25-meter dish at the National Radio Astronomy Observatory in Green Bank at two nearby, Sun-like stars: Epsilon Eridani and Tau Ceti.

This experiment was named Project Ozma after the ruler of L. Frank Baum's fictional land of Oz. Even though Tau Ceti is now known to have an orbiting exoplanet in the habitable zone, Drake saw no artificial signals in his brief experiment.

In 1961, Drake hosted a small meeting at the Green Bank radio observatory. He recalled: "I realized a few day[s] ahead of time we needed an agenda. And so I wrote down all the things you needed to know to predict how hard it's going to be to detect extraterrestrial life. And looking at them it became pretty evident that if you multiplied all these together, you got a number, N, which is the number of detectable civilizations in our galaxy."[22] In its original formulation, N is the product of these factors: the average rate of star formation in the Milky Way galaxy, the fraction of those stars that have planets, the average number of those planets that can support life, the fraction of habitable planets that actually develop life, the fraction of planets with life that evolve intelligent life (i.e., civilizations), the fraction of those civilizations that are detectable from space, and the span of time the civilizations are in a detectable or communicable state.

The Drake equation is quite a mouthful, but it's proved to be a durable way to encapsulate SETI. The first three factors are now measured by astronomers. However, the last four terms are unknown, and estimates range over orders of magnitude. Unfortunately, N is as uncertain as its most uncertain factor. Even Frank Drake agrees that his brainchild is more of a container for ignorance than a useful tool (Figure 44).

Uncertainty, however, has not deterred SETI researchers. Congress squashed NASA's funding for SETI on the grounds that it was a frivolous use of taxpayers' money, but many groups around the world are continuing the search. At the epicenter of the effort is the SETI Institute, founded in 1984 as a California nonprofit organization. It is building the Allen Telescope Array, an array of 350 antennas northeast of San Francisco, partially funded by Paul Allen, cofounder of Microsoft. SETI faces a classic "needle in a haystack" problem—that is, if

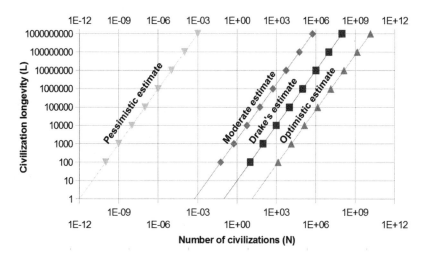

Figure 44. The Drake equation is a series of factors that estimates the number of currently communicable civilizations in the galaxy. Our degree of isolation depends on the duration or survival of civilizations for long spans of time. On the axis labels, E represents exponential notation; so, for example, 1E-03 is 0.001 and 1E+03 is 1000.

these particular needles even exist. The haystack incorporates millions of potential targets, billions of potential frequencies, and many possible ways of filtering and detecting a signal. After more than half a century of failure, it seems quixotic to continue looking, but SETI researchers point out that they are riding the wave of exponentially increasing computer power and detector bandwidth. In its first few months online, the fully built Allen Telescope Array will eclipse the sum of all previous searches.[23]

It will be easier to decide that a radio signal is artificial in origin than to decode the meaning in the signal. If you doubt that, recall that we can't communicate with primates that share 99 percent of our DNA. Now imagine that we're trying to communicate with aliens who might not even have DNA, aliens of unknown function and form. We assume that if they send radio signals, they must be intelligent. In other words, the choice of a means of communication is very telling. The medium is the message.

The main approach to SETI continues to be radio astronomy. But the power of modern lasers suggests an alternative. If a civilization on a planet was using rapidly pulsed lasers to send signals rather than radio transmitters, the pulses might stand out against the steady light of the nearby star. Optical SETI can be done with small telescopes if the stars are nearby. Pulsed lasers are now powerful enough to match the Sun's energy—but only for radiation traveling in one direction and only for a billionth of a second.

Another strategy is to look for energy used by a civilization in the form of heat "waste." Infrared telescopes can look for an excess of cool or low energy radiation, above what would be produced naturally by a star and its surrounding planets. This offers the prospect of detecting civilizations even if they are making no active attempt to communicate.

SETI is more accurately called a search for extraterrestrial technology, since it's possible to have intelligence without technology. Consider the orca. Often called killer whales, orcas are actually relatives of dolphins. They grow up to 30 feet long, weigh up to 11 tons, and can live as long as humans. Lori Marino at Emory University has analyzed an orca brain with magnetic resonance imaging (MRI), finding that their brains are large and extremely well wired for analyzing their three-dimensional environment.[24] Orcas have a complex language with dialects that vary regionally; they have dynamic social groups; they spend a lot of time socializing their young; and they seem to have hunting methods that pass from generation to generation. This ability to transmit cultural information puts them in an elite category, one previously thought to include just us. With no natural enemies but man, they're perfectly adapted to their aquatic environment and have no need to evolve fingers and opposable thumbs. Orcas will never do SETI, and SETI will never find creatures like orcas elsewhere.

Even though light is fleet-footed, space is vast. The nearest Earth-like planet is likely to be dozens of light years away. And if technological civilizations are rare outcomes of biological evolution, our nearest "pen pals" might be hundreds or thousands of light years away. A civilization might have decayed or died by the time we received its signals. As we contemplate leaving our home planet, we must prepare for the fact that space might be a very lonely place.

Given the dichotomy between being "alone" and "not alone" in the galaxy, each possibility affects our rationale for exploring deep space. If we're alone, the only reasons to venture beyond the Solar System are curiosity or a desire to propagate our civilization beyond the home planet. If we're not alone, interstellar travel is an attempt to be part of something much larger than our planet and our species.

11

Living Off-Earth

Biosphere 3.0

It seemed like a science fiction reality show on steroids. In 1991, eight men and women were sealed into a three-acre glass-and-steel complex in the Arizona desert called Biosphere 2 (Figure 45). Their mission was to live in a self-sustaining environment for two years, as a prototype of how humans might one day live on Mars, or in space.[1]

Texas billionaire Ed Bass sank $150 million into the project, and it was variously characterized in the press as a utopian dream or a rich man's folly. The occupants wore jumpsuits out of *Star Trek*—which, depending on your point of view, made them look like either consummate professionals or inmates at a county jail. Few had serious scientific credentials. The soaring architecture was inspired by Buckminster Fuller's geodesic domes, but there was also a darker backstory associated with founder John Allen, who ran a commune in New Mexico that had the trappings of a cult. Allen was a metallurgist and Harvard MBA who experimented with peyote and spent the late 1960s lecturing in San Francisco's Haight-Ashbury district. In 1974, when young Yale dropout Ed Bass arrived at Allen's Synergia Ranch, the two men instantly hit it

Figure 45. Biosphere 2 has been owned and operated by the University of Arizona since 2011. It's located a half hour north of Tucson in the foothills of the Catalina Mountains. After a checkered history as a sealed ecosystem, it operates as an Earth systems science research facility.

off, based on their shared interest in the environment. Allen had big ideas and Bass was heir to an oil fortune, so they built an 82-foot sailboat and traveled around the world studying ecosystems and sustainable development. Allen became obsessed with space colonization.[2]

As the project unraveled, former Biospherians claimed that Allen had exercised oppressive control over them, creating paranoia and low morale. Early on, Jane Poynter sliced off the tip of her finger and had to leave to get medical attention. She returned with two mysterious duffel bags that critics claimed were full of supplies. Over the first year, the occupants lost 10 percent of their body weight and had to use imported food. A carbon dioxide scrubber was installed to stop the gas from rising to dangerous levels. There were rumors of squabbling and warring factions.[3]

A Noah's ark of 3,000 animals entered the dome along with the eight humans. It included thirty-five hens and three roosters, four pygmy goats and one billy goat, two sows and a feral pig, and some tilapia fish. But due to the fluctuating carbon dioxide levels, most of the vertebrates and all of the pollinating insects died. Morning glories choked the rainforest. The cockroach population exploded. Then there were the ants. A species called *Paratrechina longicornis*, or crazy ant, killed off the other ants as well as the grasshoppers and crickets. Relentlessly, they took over the food web. Once, in a frenzy, they overran an ecologist who was one of the occupants in the dome.

It got worse. After sixteen months, the oxygen levels had declined by

25 percent, equivalent to a level available at an altitude of 13,500 feet. So oxygen was injected into the habitat, which removed any remaining pretense that it was a sealed and self-contained environment. As the first wave of occupants emerged and a new wave prepared to enter, simmering conflicts burst into the open. In April 1994, the on-site management team was ousted after being served with a restraining order by federal marshals. A few days later, two members of the first crew, Mark Van Thillo and Abigail Alling, allegedly vandalized the project by breaking panes of glass and opening an air lock and three emergency exits.[4] Two members of the second crew had to be replaced. The second mission ended prematurely after only six months.

Two decades later, it's possible to make a more balanced judgment on Biosphere 2. Anyone reading about it in the popular media would have seen grandiose expectations and hype, followed by damning criticism. The project failed as a prototype for a completely sealed environment, but more than 200 published papers have been based on research done in Biosphere 2.[5]

It was the largest "closed system" habitat ever built, with five distinct biomes: ocean with coral reef, mangrove wetland, tropical rainforest, savanna grassland, and fog desert. The problems with carbon dioxide and oxygen have been well documented, but the seal system and "lungs" that allowed the system to respond to temperature variations are the best ever constructed. Biosphere leaked about 10 percent of its oxygen per year, whereas the Space Shuttle leaked 2 percent per day. And even though the Biospherians might have smuggled in some snacks, their half-acre of land was the most productive agricultural experiment in history. The occupants got 85 percent of their food from bananas, sweet potatoes, rice, beets, peanuts, and wheat grown in the dome. They lost weight in the first year but gained back much of it in the second year as their bodies adjusted to the low-calorie, nutrient-dense diet. Most of them emerged with improvements in cholesterol level, blood pressure, and immune-system indicators.[6]

Most of what we now know about the effect of ocean acidification on coral reefs was learned in the Biosphere.[7] Wastewater was successfully treated in the artificial wetlands. And even though a few species ran amok for short periods, the overall food web stayed in reasonable balance. Before the Biosphere was built, many ecologists thought the experiment was so complex that it would be a catastrophic failure. In practice, scientists are continuing to learn through tightly controlled variables how earth systems respond to environmental change.

Despite its limitations, Biosphere provided essential lessons for anyone designing a truly sealed and self-contained environment on the Moon or Mars.[8] A high-fidelity, end-to-end, full-duration mission simulation is required before sending people off-Earth. The Biospherians could in the end just open the door and go home—Moon and Mars colonists will have very few options. The problems of food production and oxygen loss are not inherent to bioregenerative systems; they were specific to the Biosphere design and can be corrected. Most of the problems encountered were unforeseen and some were unforeseeable. Complex, miniature ecosystems are subject to nonlinear effects that compound over time. Call it the butterfly effect.

Since colonists won't be able to live exclusively in a bubble, another crucial piece of equipment is a spacesuit. Spacesuits have changed very little since the 1960s; the Americans, Russians, and Chinese all use bulky and clunky suits that offer safety but limited mobility.[9] A spacesuit has to deal with vacuum and temperature extremes; it has to protect against micrometeorites and infiltration by dust; it has to provide breathable air; and it has to monitor the occupant's vital signs. The private space industry is hiring top designers for a new generation of spacesuits; in response, NASA turned to social media by having the public vote among the final designs for its next spacesuit. The winner, called Z-2, has collapsing pleats and electroluminescent blue patches and looks strangely retro.[10]

Beyond the style makeover, there are more substantial improve-

Figure 46. In the early 1970s, NASA's ambition and vision were fully intact. At that time, the agency commissioned artwork depicting toroidal space colonies with artificial gravity holding 10,000 people. The cost of such a facility would be literally astronomical.

ments. The Apollo-era life-support system will be replaced by twin pads of absorbing material on the back of the spacesuit. One absorbs water vapor and carbon dioxide while the other vents these waste products into space. Then they switch roles. The electronic controls and power supply are much smaller and easier to replace. The back of the Z-2 suit can be attached and sealed to the spacecraft, allowing astronauts to climb out through what's called a suitport. This avoids the need for complex air locks and facilitates much longer times outside a spacecraft or a bubble dome.

In the early 1970s, NASA hired Princeton physicist Gerard O'Neill to design orbiting colonies that could support thousands of people (Figure 46). The artistic visions of these vast spinning wheels, with their interior towns and lakes and beaches and fantastic views, are breathtaking, but they're so far beyond our current capabilities that they smack of science fiction.[11] A few years ago, British designer Phil Pauley released a proposal for Sub-Biosphere 2, an undersea facility with eight habitats. While he waits for funding, he's building a rainforest biome in Saudi Arabia. Otherwise, there's no sign of a serious follow-up to the intriguing but flawed experiment in the Arizona desert.

This is a shame, because more research is needed not only to find out how we could live off-Earth, but also how we can live harmoniously on our own bruised planet.

Space Societies

Let's start by stating the obvious: It's far easier and cheaper to fix the problems of this planet than to find a way to live off-Earth.

What are the challenges that might make us want to find a new home in space? The ultimate demise of Earth will occur in four billion years when the Sun runs out of its nuclear fuel. At that point, the Sun's core will collapse and the star's violent reconfiguration will eject a layer of gas that will engulf the Earth and cook the biosphere. But long before that, the Sun will start to burn hotter as it consumes its hydrogen; about half a billion years from now, the temperature on Earth will have risen enough to make the oceans boil.[12]

Those timescales are long enough that we might be forgiven for not getting too worried. The best metric for proximate danger is the *Bulletin of the Atomic Scientists*. Starting in 1947, a group of scientists and engineers created the Doomsday Clock to show how far we were from apocalypse. As the threat of nuclear holocaust receded, the proximity of the clock to midnight started to take into account the possibility that through climate change, biotechnology, and/or cyber-technology we could cause irrevocable harm to our way of life and the planet. The clock sat at two minutes to midnight in 1953, at the nadir of the Cold War. In 1991, it receded to seventeen minutes to midnight with the fall of the Soviet Union. In 2012, however, it read five minutes to midnight because of a surge of nuclear weapons in the hands of small, unstable countries, and the sense that climate change may have passed a tipping point.[13]

Many voices have weighed in on the subject of leaving the Earth. Carl Sagan put it this way: "Since, in the long run, every planetary civilization will be endangered by impacts from space, every surviving civilization is obliged to become spacefaring—not from exploratory or romantic zeal, but for the most practical reason imaginable: staying alive." Science fiction writer Larry Niven was more succinct: "The dinosaurs became extinct because they didn't have a space program." We may be able to fend off impacts from space, but physicist Stephen Hawking sounds the alarm about other threats: "It will be difficult enough to avoid disaster in the next hundred years, let alone the next thousand or million. Our only chance of long-term survival is not to remain inward-looking on planet Earth, but to spread out into space."[14]

A mass exodus from Earth is implausible. After all, it costs $50 billion just to send a dozen people to the Moon for a few days. Elon Musk may claim he'll reduce the price of a trip to Mars to $500,000, which is a hundred thousand times less, but that seems unlikely at the moment. If the Earth becomes contaminated or inhospitable, we'll have to live in bubble domes, fix it, or suffer through it. Nonetheless, in this century a first cohort of adventurous humans will probably cut the umbilical and live off-Earth. What issues will they face?

Beyond survival, their first issue is their legal status. As we've seen, the 1967 Outer Space Treaty addresses ownership. According to Article II, "Outer space, including the Moon and other celestial bodies, is not subject to national appropriation by claim of sovereignty, by means of use or occupation, or by any other means." That seems transparent, but it doesn't mention the rights of individuals. Bas Lansdorp, the CEO of Mars One, said his legal experts looked into the treaty. He thinks that "what goes for governments also goes for individuals in those governments." If Mars One achieves its goal, thirty people will settle the red planet by 2023; the gradually expanding settlement will use more and more Martian land. Lansdorp insists that their goal isn't ownership. "It is allowed to use land, just not to say that you own it," he says. "It is also allowed to use resources that you need for your mission. Don't forget that a lot of these rules were made long ago, when a human mission to Mars was not within reach."[15]

Some space players claim altruistic motives, but none of them can succeed without revenue to fuel their dreams. What happens when profit is the only goal?

Large multinational corporations are bound by international trade law, but they could plausibly argue that they have the right to use, even to exhaust, the resources of an extraterrestrial body. A government that wanted to appropriate land on the Moon or Mars might withdraw from the Outer Space Treaty, and it's unlikely it would suffer any serious consequences. Even Mars One exists in a legal limbo. Bas Lansdorp needs to fund his $6 billion mission: "Imagine how many people would be interested in a grain of sand from the New World!"

At some point, the debate will stop being hypothetical. The history of colonization of the Earth shows that a claim of ownership is irresistible. Each succeeding generation of settlers who are born and die beyond Earth will feel less connection to the home planet. They are likely to chafe at the rules and regulations imposed from afar. Tanja Masson-Zwaan, deputy director of the International Institute of Air and Space Law and a legal adviser to Mars One, says, "I assume at some point these settlers will become more detached from Earth, and will live by their own rules."

The historical example of Manifest Destiny is misleading in the context of space colonization. Countries have grown and gained resources on Earth by seizing territory and displacing or subjugating the original inhabitants. Even in the twenty-first century, the stains of this brutal history persist. Space is a new resource. The people who leave Earth won't be taking land from anyone.[16] Eventually, they'll have to make everything they need to survive and prosper. They will create their own wealth. It will be hard to hold them to any Earth-centric legal framework if they want to be independent.

Colonization implies replacement and growth. A Mars colony can be augmented by new arrivals, but a healthy, normal culture centers on the family unit. There will be sex and there will be babies.

Sex in space hasn't progressed beyond snickering and titillation. It's the stuff of urban, orbital legend. Every couple of years, NASA and its Russian counterpart wearily deny that astronauts have had sex. The astronauts themselves stay tight-lipped. Official policy forbids it. Zero-gravity sex is tricky for several reasons. Blood flow doesn't work as well as on Earth, so men will have trouble getting erections. Sweat piles up in layers, making intimacy less pleasant. Physics is also an obstacle: The slightest push sets an object in motion. NASA astronaut Karen Nyberg once demonstrated this by using a single strand of her hair to propel herself across the cabin. Straps and harnesses would also have to be used. Given human ingenuity and desire, though, it's possible that intercourse has taken place in some quiet, dark corner of the International Space Station. But it's not written in any mission log.

Martian sex presents fewer obstacles. The 40 percent gravity would require minor adjustments. To finesse the issue of procreation, if not coupling, all-male or all-female crews have been proposed. More controversially, voluntary sterilization has been suggested for the first colonists. Mars One plans to arm its colonists with contraceptives, but it's not known how well they would work on Mars. Norbert Kraft, the medical director of the project, isn't entirely reassuring when he says they will "make colonists aware of the risks associated with having sex." The first waves of Mars colonists will die there, and they know that the medical facilities will be rudimentary; they're unlikely to want babies. But as colonies get established, the dictates of biology and human culture will prevail.

Even if we discount Mars One's plans as fantastical and hopelessly ambitious, colonization is likely eventually because there are enough pioneers with financial backing to make it happen.

When a small group of humans branch out from the root of the tree, who will they become?

Evolutionary Divergence

Imagine when the first baby is born off-Earth. That event will be an extraordinary milestone, resetting the clock of human existence. In Arthur C. Clarke's short story "Out of the Cradle, Endlessly Orbiting," an engineer at a Moon base is preparing to relocate to Mars when his wife goes into labor. The baby's first, plangent cry shakes him to his core, resonating more than the roar of any rocket ship.[17]

How many people does it take to start over? In conservation biology and ecology, there's a term called *minimum viable population.* This is the lower boundary on the population of a species in the wild such that it can survive natural disasters and demographic and genetic variations. In animal population studies, about 500 adults are required to avoid inbreeding, and 5,000 adults are required to allow a species to pursue a typical evolutionary lifespan from origination to extinction of one to ten million years.[18] These are rough estimates, used in biology to estimate the probability of extinction; in the United States, models of minimum viable population trigger protection by the 1973 Endangered Species Act.

For humans, the minimum number can be relevant during a dramatic population bottleneck. If a species population is reduced by environmental catastrophe, the genetic diversity in the remaining individuals is also reduced, and it can only grow slowly by random mutations. The robustness of the remaining population is weakened, making them more vulnerable to another adverse event.[19] This is true even though the survivors may have been the fittest individuals. Also, inbreeding is more likely, with offspring having an increased chance of recessive or deleterious traits.[20]

When geneticists sequenced the DNA of chimps and humans, they made the staggering discovery that a single band of thirty to eighty chimps can have more genetic diversity than all seven billion humans alive today.[21] We have very little genetic diversity, even though it could have developed since we diverged from chimps six million years ago. Research on mankind's restricted gene variation indicates that humans migrated out of Africa about 60,000 years ago, and at some stage before that our numbers may have dwindled to as low as two thousand. Some geneticists hypothesize that this bottleneck was caused by the explosion of the Toba supervolcano in Indonesia and resulting major environmental change.[22] Regardless of the cause, our genetic makeup hints at the fact that we were once in a perilous state, at the edge of extinction.[23]

More recent human history gives better examples of how to define the viable size of a space colony. When a new population is established by a small number of individuals from a larger population, it's subject to the *founder effect*, first described by evolutionary biologist Ernst Mayr. The founder effect leads to both loss of genetic variation and genetic divergence from the original population.

In 1790, Fletcher Christian and eight other mutineers from HMS *Bounty* were joined by twelve Polynesian women to settle on Pitcairn Island, a windswept volcanic outcrop in the South Pacific. The fifty current residents of the island are all descended from these few "founders." In 1814, fifteen British voyagers settled the remote island of Tristan da Cunha, located in the Atlantic midway between South Africa and South America. The population had grown to 300 by 1961, when a volcano erupted and everyone was evacuated to England. These small populations left the inhabitants subject to genetic abnormalities. On Pitcairn Island, Fletcher Christian spread a gene that contributes to Parkinson's disease, while the current inhabitants of Tristan da Cunha have ten times the normal incidence of a degenerative eye condition that leads to blindness.

But you don't have to be stuck on an island or Mars to suffer genetic isolation. The 18,000 Old Order Amish of Lancaster, Pennsylvania, are descended from a few dozen individuals who emigrated from Germany in the early 1700s. It's tragic that babies born into this community have a high incidence of an extremely rare and fatal genetic disorder called microcephaly.[24]

The sweet spot for a space colony may be the size of a small village. John Moore, an anthropologist from the University of Florida, developed simulation software for analyzing the viability of small groups.[25] He suggests that the optimum number for a viable long-term colony is 160. This number could be reduced with judicious genetic selection to minimize the probability of inbreeding.

If space colonists don't get "new blood" from the home planet, their gene pool will experience *genetic drift*—the change in frequency of gene variants or alleles due to random sampling. The effect is larger in smaller populations, and it acts to reduce genetic variation, which in turn reduces a population's ability to respond to new selective pressures. This may sound bad, but genetic drift and the founder effect on Earth are major drivers of evolution. They lead to the formation of new species.

Over generations, the colonists will evolve. We can imagine some of the changes that will take place. The lower gravity on Mars will alter the cardiovascular system and reduce the cross-sectional area of load-bearing bones and tendons. There will be accelerated trends in human evolution on Earth—toward being taller and having less body hair, weaker muscles, and smaller teeth. The lack of a varied natural environment will probably lead to weaker immune systems. An additional challenge will be to maintain sensory stimulation as well as intellectual stimulation, to keep the brain sharp.[26]

A new species will have evolved if off-Earth humans can no longer mate and produce viable offspring with those who never left Earth. We know this will take a long time, because a small group of people went on a one-way trip to the Americas about 14,000 years ago, and when Europeans encountered them 500 years ago, they were still the same species. Some groups in Australia and Papua New Guinea have been mostly isolated for 30,000 years and speciation didn't occur. But for colonists on the Moon or Mars, the process will be accelerated by the different physical environment and the higher incidence of mutations due to cosmic rays.

Finally, after hundreds of thousands of years and thousands of generations, when the first off-Earth baby's cry is no more than an ancestral memory, the colonists will have come of age. They will no longer be us. Imagine that the colonists live in total isolation and one day we encounter the ancestors of the people who left our planet. They'll speak their own language, have their own culture, and resemble us only partly. For each side, it will be like looking in an eerily distorted mirror.

Our Cyborg Future

It's one of the classic scenes in movie science fiction. In the cult film *Blade Runner*, the replicant Roy Batty saves "blade runner" Rick Deckard from slipping off the edge of a tall building. With superhuman strength, Batty tosses Deckard onto the roof. He then sits cross-legged and waits for his preprogrammed four-year lifespan to expire. He says to Deckard: "I've seen things you people wouldn't believe. . . . Attack ships on fire off the shoulder of Orion. I watched c-beams glitter in the dark near the Tannhäuser gate. All those moments will be lost in time . . . like tears in rain. Time . . . to die."

Roy Batty is a cyborg, as originally imagined by Philip K. Dick in his novel *Do Androids Dream of Electric Sheep?*[27] The term *cyborg*—short for cybernetic organism—was coined by Manfred Clynes and Nathan Kline in 1960. Clynes was a gifted pianist and inventor who worked as a chief research scientist at Rockland State Hospital in Orangeburg, New York; his boss was Kline, a medical researcher with more than 500 publications. They envisaged that an intimate relationship between humans and machines might help explore the new frontier of space: "Altering man's bodily functions to meet the requirements of extraterrestrial environments would be more logical than providing an Earthly environment for him in space. . . . Artifact-organism systems which would extend man's unconscious, self-regulatory controls are one possibility."[28]

Although cyborgs are the stuff of dystopian science fiction, we creep ever closer to the merger of flesh and machine.

Replacing body parts such as hearts and arms and legs has been routine for years, but cyborgs imply enhanced capabilities not present in the original human. Conventional medicine is already exploring this terrain—robotic limbs can be more powerful and flexible than the original limb, and cochlear implants can perceive sounds inaudible to a normal person. (We've already met a modern-day cyborg in the form of Tony Stark, aka Elon Musk.) Brain-computer interfaces give direct communication from the brain to an external device. They are being used to restore sight to blind people and mobility to people who are paralyzed. NASA has developed the X1 Robotic Exoskeleton to enhance the capabilities of astronauts—Iron Man is becoming a reality (Figure 47).

Neil Harbisson is a British artist born without the ability to sense color. In 2004, he started wearing a head-mounted "eyeborg," a device that converts colors into vibrations that Harbisson hears through the bones in his head. The eyeborg is referred to in his passport, making him the first government-sanctioned cyborg. The camera extends his senses by letting him hear infrasound and ultrasound, and see infrared and ultraviolet colors beyond the range of normal human vision. He

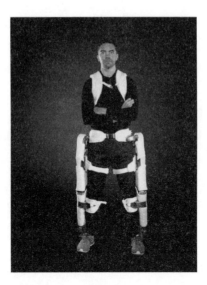

Figure 47. NASA project engineer Roger Rovekamp models the X1 Robotic Exoskeleton. It was created in the Advanced Robotic Development Lab at Johnson Space Center, to augment the capabilities of astronauts.

wants to have the device surgically and permanently attached to his skull, and he's described how the software and his brain united to give him an extra sense.[29]

Cyborgs resonate in modern culture, embodying the tension between free will and mechanical determinism. They're reminiscent of Mary Shelley's dark vision of Frankenstein, animated by electricity and overpowering its creator.

The acceptable face of cyborg research is represented by Kevin Warwick, a professor of cybernetics at the University of Reading in England. He was one of the first to experiment with implants, having an RFID chip put into his arm in 1998. Four years later, he had an array of a hundred electrodes grafted onto the nerves of his arm. This allows him to extend his nervous system over the Internet and control a robotic hand at a remote location. Warwick's wife also had a cybernetic implant, and when someone grasped her hand, he was able to feel the same sensation in his hand on the other side of the Atlantic—a bizarre form of cybernetic telepathy. "Jamming stuff into your body, merging machines with nerves and your brain, it's brand new," according to Warwick. "It's like this last, unexplored continent staring us in the face."[30]

Cyborg technology can be found in research labs but it's also gone underground. When Warwick gets an implant, he employs a team of trained surgeons; Lepht Anonym settles for a potato peeler and a bottle of vodka. She's one of a growing number of biohackers, also called grinders, who do their own implants. As she puts it, "I'm sort of inured to pain at this point. Anesthetic is illegal for people like me, so we learn to live without it." Her YouTube videos establish her as the young face of the biohacking movement. To the underground cyberhackers, computers are hardware, apps are software, and humans are wetware. One popular starting point is to have a powerful rare-earth magnet inserted into the fingertip. This lets someone sense a variety of electromagnetic fields, in addition to subways passing underground and power lines hanging overhead. Once they learn how to miniaturize them, biohackers will implant themselves with medical sensors that can talk to a smartphone and a device that will let fingers "see" by echolocation.[31] This goes beyond sensory extension to the creation of entirely new senses.

The philosophical movement that forms an umbrella for cybernetics and cyborgs is called *transhumanism*. Transhumanism is a worldwide cultural and intellectual movement that seeks to use technology to improve the human condition. Radical life extension is one aspect, as is the enhancement of physical and mental capabilities. Two prominent transhumanists are Nick Bostrom, a University of Oxford philosopher who has assessed various risks to the long-term survival of humanity, and Ray Kurzweil, the engineer and inventor who popularized the idea of the singularity, a time in the not-too-distant future when technology will enable us to transcend our physical limitations. This isn't a prospect that leaves people apathetic. Author Francis Fukuyama called transhumanism "among the world's most dangerous ideas," while author Ronald Bailey said it's a "movement that epitomizes the most daring, courageous, imaginative, and idealistic aspirations of humanity." Kevin

Warwick is committed to the cause of transhumanism: "There is no way I want to stay a mere human."[32]

Transhumanism could revolutionize space exploration. If we follow the route of nanotechnology, space probes will be miniaturized and the lower costs of manufacture and propulsion will allow us to explore a broad new range of venues in the Solar System. Alternatively, we can use robots as proxies while we're comfy in a control room on Earth. More radically, we might embrace the future seen in *Blade Runner*, where cyborgs are sent out to explore and toil. They're imbued with artificial intelligence and superhuman powers, and they have a "kill switch" in case something goes wrong. Cyborgs could be our metaphorical children—the descendants of our species—spreading out into the cosmos long after we cease to exist.

PART IV

BEYOND

Milky white light and enormous weariness. Every part of me aches but I'm immobilized. My finger twitches uncontrollably, as if it belongs to someone else. Slowly, inexorably, I become aware of my arms, my legs, my skin, as if the feeling is being retrieved from the bottom of a deep well.

My eyes flutter open. The panel just above my head registers my vital signs. In sealed beryllium coffins alongside me, one hundred fellow travelers are also stirring from a sleep close to death. Another panel shows the location of the ark. It's in a stable orbit around Proxima Centauri B1. Although I push the thought away, it muscles into my head: I am twenty-five trillion miles from home.

There's much to do. We're purposeful and don't talk much. Everyone is trying not to think about the eighteen stations where, when the lid slid open, it revealed a body that hadn't revived, one cold to the touch. A second blow came soon after. Telemetry on Ark 3 showed that it passed clear through the Proxima Centauri system. It's heading for the void of interstellar space. A meteor impact compromised the solar sail and there was no way to apply the brakes. I shudder at the thought.

Can eighty people start a new world?

Gradually, camaraderie returns and our spirits lift. We tell jokes at mealtimes and tease each other. We've traveled like flotsam for an implausible distance, a thousand times farther than anyone ever has before. I pause in front of the only window on the ark and stare out. The Sun is somewhere in that field of stars, like a buttercup set on velvet, but I can't find it. No matter what happens next, we can have pride in the achievement and the adventure. But any of us who says they're not afraid is lying.

Milky white light again. We're in the shuttle pod buffeting through the atmosphere to scout out a landing site. From Earth, our new home is a pale dot. Previous remote sensing had showed it's a living world, its air charged by photosynthesis, but we arrive knowing little about a place where we will live and die. Our mission is a huge, expensive gamble, a step across the void, hoping to find safety on the other side.

The six of us cast nervous, sidelong glances at each other. The pilot stares intently at the screen. Below us is tortured, vertiginous, and unfamiliar terrain—there's no reassuring plain or prairie, nothing like a savanna, no endless vista.

Finally, a glimpse of land through swirling clouds. Deceleration. A jolt. We don our suits and enter the air lock, as excited as children about to explore a secret garden.

It's difficult to describe the indescribable. We've landed in a verdant valley flanked by steep cliffs. Vines cling to every surface. Water drips from the cliff tops, which are partially obscured by thick clouds. There's a dense mat of vegetation underfoot. We see many plants but no animals. Everything is strange and off-kilter: gravity is weaker than Earth so I have a spring in my step, but the air is thicker so I fight back a smothering sensation. We all wear scrubber masks to keep the air breathable and filter out microbes that may be hazardous. Instinctively, everyone stays close to the lander.

Is this a swamp or Shangri-la? Either way, there's no turning back.

Working efficiently, we unpack the habitat. At the touch of a button, the memory film made of carbon nanotubes unfolds and inflates into a dome that soars twenty feet above our heads. After installing two air locks, we spend the rest of the day setting up a living space. Over the next week, the rest of the crew will join us on the surface, leaving our ark an empty, orbiting hulk, incapable of any more voyages.

Overwhelmed, exhilarated, anxious. Emotions war inside me. It seems strange to want to be alone since, as a group, we are so utterly alone. But at a break in our construction, I wander away from the dome. The only way to walk through the convoluted landscape is to follow a small stream. Looking around, I notice something strange. There are no trees. I bend down to scoop up some pebbles and rocks. Their mineral forms are familiar and reassuring; at least geology is universal.

Out of the corner of my eye, movement. I look more closely. What I thought was a mat of moss is actually a delicate web of tendrils that are moving and growing. They undulate and turn, like a carpet that's weaving itself. It seems chaotic, but suddenly the tendrils form spirals and complex geometric shapes. Then, just as suddenly, the patterns disappear. I stare, transfixed.

12

Journey to the Stars

Home Away from Home

"Prediction is very difficult, especially about the future," according to Danish physicist Niels Bohr.[1] Prediction is a core part of the scientific method. At a grainy level, scientists predict the outcome of an experiment or a measurement. At a big-picture level, scientists learn about our world by extrapolating laws of nature or predicting how they will operate in unfamiliar situations.

It's easy to cherry-pick predictions that make the prognosticator look foolish in hindsight. A classic example is that of Thomas Watson, chairman of IBM, who said in 1943: "I think there is a world market for maybe five computers." Here's Ken Olsen, cofounder of Digital Equipment Corporation, in 1977: "There's no reason for any individual to have a computer in his home." There are many other such miscalculations in the world of information technology, such as the inventor of Ethernet saying the Internet would collapse and die in 1996, and the founder of YouTube saying in 2002 that his company would go nowhere because there just weren't many videos to watch.[2] For the record, in 2014 there were two billion PCs, two billion websites, and 40 billion hours of You-Tube videos watched.

Predictions on computers and information technology tend to under-estimate the rate of progress, while those on space travel tend to over-estimate it. In 1952, writer Henry Nicholas collected predictions for the year 2000, based on the "sober conclusions of our greatest scien-tists, including many of our most famous Nobel laureates."[3] They said interplanetary travel would be common, there would be multiple Moon bases, and city-size space stations would orbit the Earth. Less than ten years ago, Burt Rutan predicted that 100,000 space tourists would have flown by 2018, and we're still stuck at seven. The reason for this dichot-omy is that information technology has gained by exponential progress in miniaturizing the components that go into computers and routers and cell phones. Space travel, on the other hand, has to deal with large objects like people and stubborn laws of physics.

It might be a fool's errand, but here's an educated guess about the arc of our near future beyond the Earth.

In 2035, a vibrant commercial space industry is operating. As effi-cient, reusable orbital flight becomes routine, prices migrate from high-end tourism to adventures accessible to the middle class. There's an unsavory underbelly to go with this new capability: reality TV shows in space, garish orbital advertising, and zero-gravity sex motels.

In 2045, there are small but viable colonies on the Moon and Mars. They depend on resupply and crew rotations from the Earth, but they successfully pioneer techniques for extracting water and oxygen from the soil and living off-Earth with a small environmental footprint. Rich countries with geopolitical ambitions foot the bill.

In 2065, mining technology advances enough to harvest resources from asteroids and mineral-rich locations on the Moon. A new business model evolves for off-Earth commerce. The United Nations and other international agencies scramble to stop the new frontier from turning into a "Wild West," but claims are often settled by corporate militias.

In 2115, a cohort comes of age who were born off-Earth and who have never been home. Colonists gain a high degree of self-governance and autonomy. Off-Earth GNP rivals the GNP of the rich nations on Earth. No economic or political imperative compels us to travel beyond the Solar System, but the visionaries are compelled to try.

Where would we go? The difficulty of traveling in interstellar space will limit us to the closest habitable location. As we've seen, there's evidence for an Earth-size planet orbiting the closest Sun-like star to the Earth, Alpha Centauri B. That exoplanet is much closer to its star than Mercury is to the Sun and has a surface temperature of 1200°C—so hot that its surface would be magma. Doppler data are not currently good enough to detect Earth-like planets farther out. However, Alpha Centauri B has a binary companion, and the orbit is wide enough that it wouldn't disrupt the orbits of planets in the habitable zones, which would be 0.7 astronomical units (AU) from Alpha Centauri B and 1.3 AU from the more luminous Alpha Centauri A. The system offers a double shot at finding a habitable planet.[4]

Simulations set expectations as we wait for better data. A 2008 study looked at how planets might form from the disk of rocky material around Alpha Centauri B. The orbits of several hundred protoplanetary rocks as large as the Moon were tracked for 200 million years (which takes only a few hours on a powerful computer). Although the number and type of exoplanets formed depended on the initial conditions in the disk, on average the simulations generated twenty rocky planets, ten of which were in the star's habitable zone. Statistics should be similar for Alpha Centauri A (Figure 48).

In 2013, Antonin Gonzalez advanced this research when he estimated an "Earth similarity index" for exoplanets in the simulations. This index gauges how Earth-like a planet is—based on surface temperature, escape velocity, size, and density. Zero represents a dissimilar planet, and one would be a planet identical to the Earth. For comparison, Venus has an Earth similarity index of 0.78 (similar in size but much hotter),

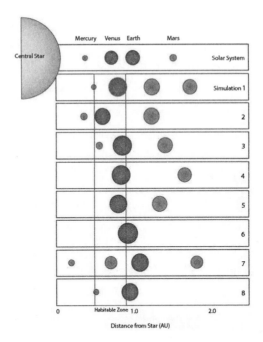

Figure 48. Results of astrophysical simulations of exoplanet formation in the Alpha Centauri system. Terrestrial planets form readily around either star, with masses and distances similar to the architecture of the inner Solar System (shown at the top for reference).

while Mars has an Earth similarity index of 0.64 (smaller and much colder).[5] The calculations assume hospitality for life as we know it. If exoplanets host biology with a radically different chemical or metabolic basis, we may not be able to recognize it, or even know how to define it.

Five of the simulated exoplanets were deemed capable of supporting photosynthetic biology. Their Earth similarity indices were 0.86, 0.87, 0.91, 0.92, and 0.93; two of them even had *better* conditions for life than the Earth.

That sounds promising, but we can't travel trillions of miles without being sure. Unless a planet has an oxygen-rich atmosphere, we'd be better off creating an artificial environment in space or even terraforming Mars. We've seen that transforming Mars to have a breathable atmosphere has been subject to feasibility studies by NASA. The technical challenges are manageable, but it would take an industrial-scale application of existing technology. The best guesses on time and cost are a millennium and a trillion dollars. The closer an exoplanet is to having Earth-like atmospheric composition, the easier the task of terraforming becomes.

To astronomers, oxygen is the best "biomarker," or tracer of life on another planet. If life on Earth is representative of biology elsewhere, low levels of oxygen indicate microbial photosynthesis and high levels are signatures of plant life. Putting it another way, if all life on Earth died overnight, the one in five oxygen molecules we breathe and depend on would disappear in a few thousand years as they reacted with rock and water. Substantial levels of atmospheric oxygen are difficult to sustain by geological processes alone. A related biomarker is ozone, which has a strong spectral signature although it's far less abundant than oxygen. Another biomarker is methane, generated by fossil fuels and decaying vegetation. Methane has low concentration today, but between 3.5 and 2.5 billion years ago it was as abundant as oxygen is today, produced by microbes called methanogens. Water vapor is also a biomarker, since we assume that life can't exist without water.[6]

In practice, astronomers will need to detect a suite of biomarkers, and compare the spectra to models of planetary chemistry and geology, before being confident they've seen biology.[7] The detection will most likely come from dispersing the feeble reflected light of an Earth-like exoplanet into a spectrum. If the light shows absorption by some combination of ozone, oxygen, methane, and water vapor, the result will make immediate headlines as the first detection of life beyond Earth. Since it will be a form of microbial life and there will be no pictures of it, public interest is likely to wane. But for science it will be a momentous discovery. Until we find the second example of life, it's always possible to argue that life on Earth is a unique accident.

The challenging observations required to detect biomarkers have only been tested on Jupiter-mass exoplanets, where no life is anticipated. In proof-of-concept measurements of six exoplanets with the Hubble Space Telescope, vapor of sodium, methane, carbon dioxide, carbon monoxide, and water has been seen.[8] Making this observation on Earth-mass exoplanets will require a new telescope in space or innovative image-sharpening methods using ground-based telescopes. Most researchers expect the critical observation to be made within the next decade. Based on the frequency of Earths found by Kepler, the nearest clone is likely to be a dozen light years away. If we get lucky, it will be even closer.

––––––––––

If we revisit the smaller-scale journey that opened this book, we see that the migration from Africa to Chile was as grandiose as reaching for the stars. The cradle of humanity was in northeastern Africa. A canny hunter-gatherer might roam 10 miles to find food. But humans migrated a distance a thousand times larger, with no certainty of food or shelter—the same ratio as the distance to nearby stars compared to the size of the Solar System.

After a thousand generations, we had traveled to the dense forests and vertiginous valleys of Southeast Asia. After two thousand more generations, we had roamed into the barren tundra of Siberia and across the land bridge to the Pacific Northwest. After another few hundred generations, we reached the lush rainforest and azure waters of the Central American isthmus. It took a scant hundred generations more to reach the southern tip of Earth's land mass. To anyone who had a cultural memory of the African savanna, the sight of the wild, windswept shore of Patagonia, with the stars in the night sky wheeling in the opposite direction, would have seemed as alien as an exoplanet.

Building a Better Engine

To see that interstellar travel is a stretch goal for space exploration, consider this scale model.

Shrink the Earth to the size of a Ping-Pong ball and the Moon would be a marble a yard away. If you hold the Earth in front of your nose and the Moon at arm's length, that's the full extent of human venturing. In this version of the Solar System, shrunk by a factor of a hundred million, the Sun would be a glowing gas ball eight feet in diameter a hundred yards away and Neptune would be the size of a beach ball four miles away. On this scale, the nearest star to the Sun is 30,000 miles from the little Ping-Pong ball, or more than the Earth's circumference. To get to the stars in a reasonable time, we need enormous speeds. It would take 50 million years to get to the Alpha Centauri system at the highway speed limit. At the speed Apollo used to get to the Moon, it would take 900,000 years, and even at the speed of the Voyager spacecraft (which left the Solar System traveling at 37,000 mph), it would take 80,000 years.

Chemical energy is just too inefficient to get us to the stars. We have to go beyond rearranging electrons among atoms and unlock the power of the atomic nucleus.

Let's revisit the energy available from different fuels. The usual units are millions of Joules per kilogram (MJ/kg). For reference, a million Joules is the energy released by a kilogram of TNT exploding, or the energy expended by running for an hour, or the energy stored in a candy bar. In these units, wood and coal store about 20 MJ/kg, gas and other hydrocarbon fuels store about 40 MJ/kg, and hydrogen has the best energy storage, at 142 MJ/kg. NASA scientists at the Glenn Research Center in Cleveland have worked out how much fuel would be required to get a Space Shuttle payload (think of a fully laden school

Energy Content in Different Fuels

| Twinkies | Coal | Gas | Deuterium/Tritium |

Figure 49. The energy storage in three different chemical fuels (food, coal, and gasoline) compared with mass-energy release in the fusion process. Matter–antimatter annihilation would be a thousand times more efficient than either fission or fusion.

bus) to Alpha Centauri in 900 years.[9] The answer is discouraging: All the mass in the universe in the form of rocket fuel couldn't do it!

If we look beyond chemical energy, the best source is mass itself. The implication of Einstein's iconic equation $E = mc^2$ is that mass is frozen energy. Because the speed of light is a large number, a tiny amount of mass converts into a huge amount of energy. Mass can be liberated into energy in nuclear reactions with an efficiency of 0.1 percent for fission, 1 percent for fusion, and 100 percent for matter–antimatter annihilation (Figure 49). That represents energy storage of 10^8 MJ/kg for fission, 10^9 MJ/kg for fusion, and 10^{11} MJ/kg for matter–antimatter annihilation. Surely an efficiency millions of times better than chemical fuels can get us to the stars?

Yes, but not without the necessary technology development. A rocket needs thrust (or force) so it can exert a big push, but it also needs a high specific impulse, which is the force delivered per kilogram of fuel per second, analogous to fuel efficiency. Chemical rockets have high thrust but lousy specific impulse. The ideal interstellar rocket must score well for both quantities. Remember the rocket equation? It says that the final speed of a rocket depends on the fuel exhaust speed and the ratio of the fuel mass to payload mass. Having a nuclear fuel means less mass is needed, but that ratio is inside a logarithm, which suppresses its influence on the final speed, so it's just as important to increase the exhaust speed.

None of the rocket engines about to be described have ever been built. They all rely on bleeding edge technology, though they are comfortably within the realm of known physics. Let's look at the potential performance of rockets that don't depend on chemical energy.

For nuclear fission, the simplest concept is to put a reactor on top of a rocket nozzle. Conventional fission and fusion concepts—recall that fusion hasn't yet been used to generate energy on Earth—have ten to twenty times better performance than chemical rockets. This is a key limitation, since the practical gain in a rocket ends up far less than the theoretical gain based on energy density.

Fusion would still require 10^{11} kilograms of fuel, equal to a thousand supertankers, to get to Alpha Centauri in less than a thousand years. In the 1960s, Project Orion was developed by Stanislaw Ulam, a brilliant mathematician who had worked on the Manhattan Project. The idea was to use a series of controlled nuclear explosions to propel the spacecraft forward. In the 1970s, the British Interplanetary Society amended the design to use a large number of microfusion explosions (Figure 50).[10]

With matter–antimatter engines, we enter the realm of speculation, since antimatter, the quantum shadow partner of matter, has only been produced and contained in tiny quantities. An antimatter engine has a

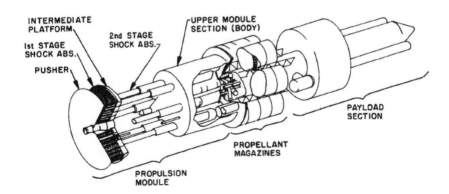

Figure 50. NASA's version of the Project Orion concept, where pulsed nuclear fusion projects the power. The design combines high thrust and high exhaust velocity. No current technology can harness nuclear explosions in this way.

performance gain of a factor of a hundred, and the fuel requirement drops to about 100,000 kilograms or ten railway tankers of propellant to get there in less than a millennium. These numbers double because the spacecraft will need fuel to decelerate when it reaches its destination. Gathering this much antimatter will be impossible for the foreseeable future. At the moment, it would cost $100 billon just to create one milligram of antimatter.[11] For those wanting to try this at home— the calculation, not actually building an interstellar rocket—the RAND Corporation used to sell the nifty (and very retro) Rocket Performance Calculator, dating from 1958.[12] This circular slide rule incorporates the rocket equation and it can still be found occasionally on eBay, making for a great conversation piece.

The energy requirement of interstellar travel is formidable. Sending a 2,000-ton, Space Shuttle–size craft to Alpha Centauri in fifty years (a tenth the speed of light) costs 7×10^{19} Joules, assuming a perfect conversion of energy into forward motion, which isn't true of any real propellant. That's the energy consumption of the entire United States for six months. If that energy came from nuclear explosions, it would take a thousand Hiroshima bombs. The energy requirement can only be reduced by having the smaller payload or by traveling more slowly and taking longer to get there.

A clever alternative avoids carrying and accelerating all that fuel.

An interstellar ramjet would employ a magnetic "scoop" a thousand kilometers across to grab protons from the near vacuum of space and fuel a nuclear reactor. The idea originated with American physicist Robert Bussard in 1960.[13] He was assistant director of the program to develop fusion power under the Atomic Energy Commission in the 1970s, and his ramjet concept was quickly coopted to become a staple of science fiction. There are huge engineering issues in realizing this concept. The physical challenge is to gather enough fuel from the sparse interstellar medium—the scoop has to sweep the equivalent of the volume of the Earth just to get one kilogram of hydrogen—while producing enough thrust to overcome the drag of the fuel collected. Slowing down at the destination is another problem, as yet unsolved.

Solar sails are still more promising. The appropriately named Robert Forward developed a similar concept in the mid-1980s where a 10 million gigawatt laser shines through a 1,000-kilometer Fresnel lens onto a 1,000-kilometer sail. Unfortunately, 10 million gigawatts is a hundred times the energy consumption of all countries on Earth. Unfazed, Forward retooled his idea into a 10 gigawatt beam of microwaves that push on a kilometer-wide grid of fine wires. His "modest" proposal could be done with the energy output of ten large electrical generating plants.[14] Forward was a dapper engineer, known for his shock of white hair, owlish glasses, and eye-popping vests. He died in 2002, but his ideas are still very influential.

In the late 1980s, Dana Andrews and Robert Zubrin came up with the concept of the magnetic sail.[15] A solar sail is driven by radiation from the Sun while a magnetic sail is driven by the solar wind, a diffuse plasma of charged particles streaming out from the Sun. The plasma would be harnessed by the magnetic field created by a large loop of superconducting wire. The magnetic sail has the disadvantage that the solar wind carries thousands of times less momentum than sunlight, but its big advantage is that the momentum is gathered by a massless magnetic field rather than a large physical sail.

An alternative to using the Sun to propel a sail is to beam the energy to it from the Earth, with quick enough acceleration that it could coast to the destination. Two heads are better than one for this idea. James Benford, president of Microwave Sciences, believes that a microwave beam is superior to a laser for accelerating a solar sail. His lab experiments show that high-intensity microwave beams could be developed, but the sail material has to be extremely light and robust and could only withstand the temperature of 2,000 degrees by being highly reflective.[16] His twin brother, Gregory Benford, is professor of physics at the University of California in Irvine and a noted science fiction writer. They collaborate on this project, and on gathering together the hard-science experts and science fiction visionaries to brainstorm the future of interstellar travel.[17]

The 100 Year Starship project is funded by NASA and the Defense Advanced Research Projects Agency (DARPA). In 2012, a million-dollar grant was awarded to former astronaut Mae Jemison and the nonprofit organization Icarus Interstellar for work toward interstellar travel in the next hundred years. It's important to realize that the majority of the speculative research on interstellar travel is being undertaken by professional physical scientists and engineers, with the work published in scholarly journals and books.[18]

Thomas Jefferson thought it would take a thousand years for the American frontier to reach the Pacific. It happened in less than one-tenth of that time. And technology advances swiftly; the first nuclear reactor in 1942 in Chicago generated half a watt, but within a year a reactor was constructed that could power a small town. In the first fifty years of its development, the most powerful laser has increased in intensity by a factor of 10^{20}. Returning to the analogy with information technology, linear progress in propulsion technology will not be sufficient to reach the stars; there will have to be technological leaps. Physicist Andreas Tziolas, president of Icarus Interstellar, says, "I have faith in our ingenuity."[19]

Here Come the Nanobots

The nearest Earth-like planet is likely to be in our cosmic backyard in a galaxy a hundred thousand light years across. Remote sensing might indicate life on that planet, but the evidence will be limited and possibly ambiguous. Going there ourselves will, for the foreseeable future, be fiendishly difficult and ruinously expensive. Costs are variable, but they run upwards of a hundred trillion dollars, the value of current world GDP. Is there another strategy?

Nanobots could reduce the cost and the energy requirements drastically. The US military's smart motes for the battlefield give a sense of the possibilities. Space researchers can piggyback on a relentless push for miniaturization that's motivated by medical applications. We imagine a fleet of baseball-size spacecraft, each crammed with sensors and a small camera, sailing toward the nearest Earth-like planet. As they arrive and drift down through the atmosphere, they transmit video back to Earth. There's redundancy, so if some are lost in transit or fail to make it to the surface, the mission isn't lost. We'd send nanobots in waves, so they could pass information back down the route of travel, like a bucket brigade at a fire. That reduces the power requirement for the transmitters on each nanobot. The mission would take a generation, but we can imagine expectation building as the fleet reaches its destination: Huge screens in city centers around the world carry the video feeds and crowds gather as the first images reveal details of an exotic new world.

Going from tons to kilograms makes everything easier, but it's not a slam dunk. Tony Dunn has crunched some numbers using solar sails for propulsion. With existing materials like Mylar, a kilogram nanobot can only reach a terminal speed of 80 kilometers per second, just five times faster than the Voyager spacecraft and hopelessly inadequate for the task. Making the sail larger than a hundred square meters means all the energy is going into accelerating the sail rather than the payload. Sail materials a million times lighter than Mylar would be needed to reach 10 percent of the speed of light. Using a laser to beam power from Earth directly to the sail helps. Now the solar sail needs to be only one meter across. The difficult trick is aiming the laser at such a small target when it's far away. At the distance of Neptune, the laser would have to be targeted 100,000 times more accurately than the Hubble Space Telescope. A readily available 30-kilowatt laser could propel the kilogram probe to

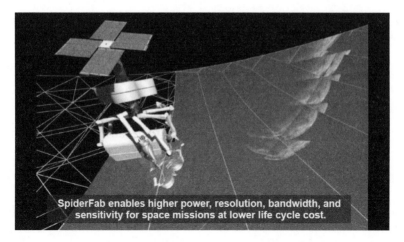

SpiderFab enables higher power, resolution, bandwidth, and sensitivity for space missions at lower life cycle cost.

Figure 51. NASA is collaborating with Tethers Unlimited on a space fabrication system. In this mockup, a space robot 3-D–prints the backbone for a mile-wide solar array. Creating structures in orbit is far cheaper than sending them there by rocket.

Alpha Centauri in forty years. The cost of the electricity: $800 million, assuming a residential rate of 15 cents per kilowatt hour. For a fleet, the price tag climbs to $100 billion. That's steep, but doable.[20]

Miniaturizing a spacecraft is a logical strategy, but it's unimaginative. Nanotechnology suggests other possibilities: self-assembly and self-replication. In 2012, a company named Tethers Unlimited won a NASA contract to develop a system called SpiderFab.[21] Spiderfab aims to use 3-D printing and robotic assembly to fabricate components in orbit—solar arrays, trusses, and shrouds that are ten times bigger than those that can currently be put in orbit (Figure 51). In the lab, self-assembling machines are showing great promise. MIT researchers have created cubes no larger than dice that hold sensors, magnets, and a tiny flywheel. Identical cubes can all be commanded to move, snap together, and form arbitrary shapes.

An even more exciting capability is self-replication. Eric Drexler talked about it in his prescient 1986 book on nanotechnology, *Engines of Creation*. Even earlier, in lectures at Princeton University, physicist Freeman Dyson described thought experiments involving large-scale replicating machines. In one, spacecraft traveled to Saturn's small moon Enceladus, mined material to replicate themselves, and also launched spacecraft powered by solar sails to carry ice to Mars and begin to terra-form the red planet. As with self-assembly, self-replication has made more progress in the lab than in space.

The RepRap Project began in 2005 with the goal of designing a 3-D printer that could create most of its own components. It started at the University of Bath in England, but the code for computer-aided design and manufacture is open source, so the project has spawned a large developer community. In 2008, the RepRap machine "Darwin" pro-duced all the parts needed to make an identical "child" machine. The project will make its technology freely available to anyone, with the goal of helping people make artifacts for everyday life.[22]

The ultimate expression of self-replication is a von Neumann probe. This is a spacecraft that could go to a neighboring star system, mine materials to create replicas of itself, and send those out to other star sys-tems. Using fairly conventional forms of propulsion, these probes could spread through a galaxy the size of the Milky Way in less than a few million years. The probes could investigate planetary systems and send information back to us on the home planet.[23]

The concept is named after the Hungarian mathematician and physicist John von Neumann. He was one of the major intellectual fig-ures of the twentieth century, making important contributions to math-ematics, physics, computer science, and economics. Noted physicist Eugene Wigner recalled that von Neumann's unusual mind was like a ". . . perfect instrument whose gears were machined to mesh accurately within a thousandth of an inch." But he was less perfect in the real world. As a driver, he had numerous accidents and a few arrests, usually because he was distracted or reading. He overate, told off-color jokes, and did his best work in noisy and chaotic environments.

In the 1940s, von Neumann figured out the logical requirements for self-replication. He described a computational "machine" that could make copies of itself, allow for errors, and evolve. This remarkable work preceded computers and anticipated the later discovery of DNA and the mechanisms of life. His work was theoretical, but it created a road-map for building actual self-replicating machines.[24]

Perhaps this is the way we will eventually explore the galaxy. Diffusing through interstellar space and exploring distant worlds with a fleet of self-replicating probes sounds fantastical, but it could be achieved with a reasonable extrapolation of our current technology. Which raises the question: Has any other civilization done this?

Warp Drives and Transporters

No fundamental obstacle prohibits the creation of a propulsion system that can accelerate a payload to a significant fraction of the speed of light. The highest speed ever reached by a spacecraft was 165,000 mph or 25 miles per second, when the probe Juno used Earth's gravity to catapult toward Jupiter. That's fifty times faster than a bullet, but only 0.01 percent of the speed of light. Reaching the nearest stars in less than fifty years would require speeds a thousand times faster, or 10 percent of the speed of light.

Let's now venture beyond the bounds of projected capabilities based on well-established science, into the realm of speculation and science fiction.[25]

Two staples of science fiction, and routine occurrences on *Star Trek*, are warp drive and teleportation. A warp drive enables faster-than-light travel. Einstein's theory of special relativity posits the speed of light as an absolute limit for the transmission of matter, energy, or information of any kind. Special relativity is a foundational principle in physics, so that would appear to kill the possibility of a warp drive. Tachyons—fundamental particles that travel faster than light—were hypothesized in 1967, but no evidence for them has ever been seen.[26] In 1994, physicist Miguel Alcubierre proposed a theoretical solution for faster-than-light travel based on negative mass.[27] The consensus among physicists is that a warp drive is not possible under the known laws of physics, but the idea got some attention at the 100 Year Starship Symposium at the Johnson Space Center in 2012.

What about teleportation? Imagine this situation. You're about to step into a device that will deconstruct your atoms into an energy pattern, beam the information to a remote target, and rematerialize you.

In "Realm of Fear," the 128th episode of the TV series *Star Trek: The Next Generation*, Lieutenant Reginald Barclay develops a fear of the transporter that's used to beam crew members down to the surface of a planet. He becomes obsessed with all the things that could go wrong when the 10^{28} atoms in his body are dismantled and then reassembled.[28] Eventually, his fear becomes debilitating.

There's no formal term for this condition.

The TV series and subsequent films were sketchy on how transporter technology works. It's supposed to transport objects accurately at the level of individual atoms, using something called a Heisenberg compensator to remove uncertainty from subatomic measurements. When technical adviser Michael Okuda was asked how it works, he said, "It works very well, thank you." On the original *Star Trek* show, the special effect for the transporter was created before computer animation existed, so it was low tech: a slow-motion camera was turned upside down and it filmed backlit grains of aluminum powder falling in front of a black background.

Classical teleportation measures every atom in the human body, encoding that information into photons, sending the photons to a remote location, and using the information to reconstruct a perfect replica of the body. That's just an engineering problem. But with 10^{28} atoms to deal with, it's a very nasty engineering problem.

For decades, it was thought that teleportation defied physics. Heisenberg's uncertainty principle says that we can't simultaneously

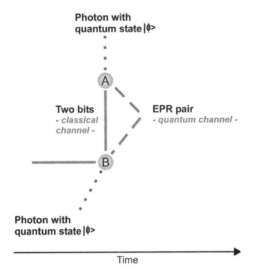

Photon with quantum state |φ>

(A)

Two bits
- classical channel -

EPR pair
- quantum channel -

(B)

Photon with quantum state |φ>

Time

Figure 52. Theoretical diagram for the quantum teleportation of a photon. In this Feynman diagram, two bits of information would move classically from A to B; in quantum teleportation, information is transmitted via a single entangled qubit.

and accurately measure all the properties of even a single atom, let alone vast numbers of them. Measuring any property of a subatomic particle changes its state. So there's no way to convey that state to a remote location with high fidelity.

In 1993, physicist Charles Bennett and his team made a breakthrough. They realized that particles at two different locations could be induced into something called *quantum entanglement*, where information about their physical states was shared. The loophole that lets us circumvent Heisenberg's uncertainty principle involves trying not to know too much. We disturb the particle before we measure it, so we never know its state. Then we can subtract that disturbance at the other end to re-create the original state of the particle (Figure 52).[29] Think of entanglement as a black box that conceals but connects events at two locations remote from each other. It seems to violate causality because changes at the two locations occur instantly, but there's a limit to what we can know or measure. Quantum entanglement has been demonstrated using photons, electrons, buckyballs, and even small diamonds. It's pure quantum weirdness.

Let's personalize it. Alice wants to teleport something to Bob.[30] An entangled pair of photons serves as the intermediary in their experiment. She measures a property of her photon where the outcome depends on the entangled state of the pair. She records her measurement and sends it to Bob. He can't tell what the state of her photon was, because the entanglement used in the measurement hides the true nature of that state. What Bob can do, however, is use information from Alice to modify the state of his photon. Then he can re-create the exact state of the photon Alice originally measured.

Even though the entangled state spans two separate locations, Bob can't complete the teleportation until she sends him the result of her measurement. So the special theory of relativity and causality aren't broken. The process allows information to be copied with perfect fidelity, although teleportation doesn't literally make a copy; it shifts quantum information from one place to another, destroying the original in the process.

Progress is rapid in this exciting research field. Physicists first demonstrated quantum teleportation in the lab in 1998 over a distance of a meter. In 2012, a research group teleported information between two locations in the Canary Islands 143 kilometers apart. In 2013, worldwide teleportation was demonstrated.[31] The reliability of teleportation is also improving dramatically. In 2009, transfer of quantum information over distances of a few meters succeeded only one in 100 million times. In 2014, scientists at Delft University in the Netherlands teleported the quantum state of two entangled electrons with 100 percent reliability.[32]

The mechanism of quantum entanglement is being used for cryptography and it's likely to play a role in developing faster computers, but most physicists think it's unlikely we'll ever be able to create and interrogate the quantum entangled state of more than a few thousand atoms. So transporters aren't even on the far horizon.

Star Trek's Lieutenant Barclay was only worried about his atoms being scrambled so that he turned into a pile of unrecognizable goo. But he also should have been worried about the philosophical implications of teleportation. You are not your particles. The atoms in your body fall off and get replaced all the time. Toast turns into eyelashes. You and your thoughts and your genetic information are really patterns rather than piles of particles. So when a transporter disassembles you, it kills you; when it reassembles you elsewhere, it gives birth to you. Logically it could do that as many times and at as many locations as it liked. Where would that leave your sense of self?

13

Cosmic Companionship

Number of Pen Pals

We've seen how hard it is to leave the cradle of Earth, even for a short time. However, enough ingenuity and technology are being applied to the problem that it's only a matter of time before we spread beyond our planet. This raises a series of questions. Are we the only species to travel in space? Are we the first? If we are not, how would we know? It will spur space exploration if we know we are not the first or the only species to spread our wings beyond the home planet.

The question of cosmic companionship brings us back to the Drake equation and all its embedded uncertainty. Recall that the equation is a multiplicative set of factors incorporating astronomy, biology, and sociology and designed to give an estimate of the number of civilizations that are communicating or traveling through space at a given time.

Exoplanet surveys suggest there are 10 billion Earth-like planets around Sun-like stars. That's a vast number of "Petri dishes" in the Milky Way: locations with suitable physical conditions and the chemical ingredients for biology. Scientific arguments based on a sample of one are unreliable, but the fact that life formed on Earth as soon as suitable conditions arose is taken as evidence that habitability almost always means actually inhabited. A counterargument is that life only seems to have arisen once on this planet, but that argument is weak because other origination events may have been lost or concealed or outcompeted by the existing form of life. If life is found on Mars, either present or ancient, it will be good evidence that the fraction of habitable planets that host life is close to one. If we assume that for a moment, the Drake equation becomes $N \sim f_i \times f_c \times L$. Here, N is the number of civilizations in our galaxy that are currently able to communicate through space, f_i is the fraction of planets with life that go on to develop intelligent life, f_c is the fraction of those that can communicate through space, and L is the length of time that they endure or have such a capability.

At this point, opinions diverge and uncertainty rules. Some biologists argue that f_i is low because only a handful of the hundreds of millions of species on Earth developed intelligence. Others argue that biology has trended toward greater complexity over time, and there may be an evolutionary advantage to the development of brains. The fraction f_c is even more controversial. There are intelligent species on Earth that can't send signs of their existence into space—elephants, orcas, octopuses, and others. They could have hypothetical counterparts on other worlds. There are also many reasons why a technological civilization may choose not to travel in space, communicate, or somehow reveal its existence. We lose all traction on logic when we enter the realm of alien sociology.[1]

The last term, L, is also imponderable. Anatomically modern humans originated 200,000 years ago, culture and language evolved 50,000 years ago, and the first civilizations date back to 10,000 years ago. We've had space travel and SETI for only about fifty years, which might suggest the number 50 as a lower bound on the lifetime factor, but in that technological surge we developed the ability to destroy civilization via nuclear weapons, so the lifetime may be short if technology renders a civilization unstable.[2] Carl Sagan speculated that the last few factors were close to one, meaning approximately that N equals L (etched on Frank Drake's California license plate), so civilization lifetime governs the number of potential pen pals. Sagan's view of the Drake equation spurred his strong advocacy of environmental issues and his warnings about the dangers of a nuclear holocaust.

The lifetime factor is a reminder that the universe contains real estate of time as well as space. The cosmos is 13.8 billion years old, and the Milky Way started forming soon after that. As generations of stars live and die, the galaxy increases its abundance of the heavy elements needed to make planets and biology. The galaxy disk formed nine billion years ago and Earth-like planets could have first formed then, giving them a 4.5-billion-year head start on the Earth. The Drake equation ignores the fact that civilizations could emerge more than once on a particular habitable planet. A civilization may be destroyed by disease or natural disaster or internal instability, but others might emerge over the eons. The Drake equation doesn't distinguish between active communication and the passive creation of a detectable technological footprint.

If biology is abundant but evolution rarely leads to intelligence and technology, or if technological civilizations are short-lived, then the number of pen pals in the galaxy is low. We might be truly alone. But if evolution almost inevitably leads to space travel and communication, or if technological civilizations are durable, then the Milky Way could be buzzing with activity.

Finally, the Drake equation factors in only one galaxy. The Milky Way is not unique among the hundred billion galaxies seen within the limit of modern telescopes, so we extrapolate the local census of life into the vastness of the observable universe. Even if the number of intelligent civilizations in the galaxy is only a dozen, that still projects to a trillion throughout space and time. The total number of technological civilizations in the universe could be truly staggering.

The Great Silence

Speculation is fun, but science is all about data, and SETI researchers have been "listening" for artificial radio signals from nearby stars for more than half a century. What have they heard?

Nothing. It's referred to as "The Great Silence," as if radio waves were audible or sound could travel through space. Radio astronomers have been listening for pulsed radio signals because radio waves are not produced by stars, they have low energy, and they travel easily across large distances in the galaxy. For all these reasons, it's assumed they would be the tool of choice for a technological civilization trying to communicate. The targets are relatively close Sun-like stars that might have planets around them. Stars are the targets because their attendant planets wouldn't be visible and they would be so close to the star that they couldn't be distinguished on the sky. The data are analyzed by computers, and when the radio intensity is converted into sound, all they ever hear is static. White noise. Hiss.

The epitome of SETI success was portrayed in the 1997 film *Contact*, based on Carl Sagan's 1985 science fiction novel of the same name.[3] The astronomer Ellie Arroway, played by Jodie Foster, is in the control room of the Very Large Array in Socorro, New Mexico, watching as the twenty-seven radio dishes tilt in the direction of the bright star Vega. She settles into her chair to listen as the radio signals are converted into sound in her headphones. Suddenly, pure static is interrupted by a thumping sound. Then a pause. Then two more thumps. The sounds come in clusters. Gradually it becomes obvious they're prime numbers. The premise is clear. Stars don't emit radio waves, so the radio signal must originate from technology on a nearby planet. Mathematics is assumed to be a universal language, and it's implied that it would take intelligence for any species to calculate a prime-number sequence.

Science fiction writer Ursula K. Le Guin imagined life on a planet around the star Tau Ceti (one of the two stars observed by Frank Drake in his Project Ozma in 1959).[4] That civilization built its religion around mathematics, and its denizens "chanted the primes." Arroway discovers that later signals include a huge amount of coded information, including the instructions for building a machine that can transport people through wormholes.

Reality is more mundane. No convincing signal from ET has ever been detected.

SETI began with the pioneers of radio. In 1899, Nikola Tesla observed repetitive signals in his coil transformer that he thought had originated from Mars. A few years later, Guglielmo Marconi also believed he had picked up messages from Mars. It's likely they both were witnessing natural phenomena in the Earth's atmosphere.[5] Following Project Ozma, the Soviets did pioneering work and the largest experiment in the United States was a radio telescope the size of three football fields called "Big Ear" at Ohio State University. In 1977, a technician at Big Ear saw a booming signal on the printout and annotated it with an exclamation. The "Wow!" signal never repeated and was never identified with a celestial source; scientists consider it a dead end. Radio SETI involves searching for narrow band signals, typically less than 100 Hertz wide. That's because a signal confined to a narrow slice of the radio dial indicates a purpose-built transmitter—think of your car radio scanning to find a station. Natural sources of radio waves like pulsars and quasars spread their signal over a relatively broad frequency range.

SETI has progressed in fits and starts. In 1959, Frank Drake scanned his single-channel receiver over a 400-kilohertz (kHz) band, a painfully slow way to search the spectrum. Paul Horowitz transformed the search in the 1980s. Horowitz was a wunderkind who became a ham radio operator when he was only eight. As a professor of electrical engineering and physics at Harvard, he wrote *The Art of Electronics*, considered the bible in its field. In 1981, he developed a 131,000-channel spectrum analyzer that could fit in a suitcase. By 1985, he increased that to 8.4

million channels, with funding assistance from film magnate Steven Spielberg. A decade later, a receiver equipped with custom digital signal processing boards was able to scan 250 million channels every eight seconds.

Technological breakthroughs catalyzed SETI, but political headwinds slowed the progress. In 1978, SETI received one of the infamous "Golden Fleece" awards from Senator William Proxmire. He gave the award monthly to projects he thought were egregious wastes of public money, and he dumped particular scorn on the "search for little green men." In 1981, he added a rider to the NASA budget that prevented the agency from doing SETI research. Carl Sagan persuaded Proxmire to relent, but in 1993 the program was killed again, this time by Nevada Senator Richard Bryan, who noted with satisfaction the end of a "great Martian chase" at the taxpayers' expense.[6] His action was hypocritical, since he later lobbied for government funding to upgrade a Nevada state highway that runs close to Area 51 (an iconic site for UFO conspiracy theorists) and name it the Extraterrestrial Highway. Since 1995, the program has continued with a mix of private and federal funding.

There's also been pushback against SETI within academia. Harvard biologist Ernst Mayr called SETI "hopeless" and "a waste of time," and he criticized his colleague Paul Horowitz for drawing graduate students into such an endeavor. Sagan rebutted Mayr, but the search continues to elicit strong opinions.

SETI uses both listening and signaling strategies. Inevitably, SETI is anthropocentric and its strategies are tightly coupled to our current capabilities. Its history mirrors the evolution of our technology. In 1820, German mathematician Karl Friedrich Gauss suggested cutting a right-angled triangle into the Siberian forest, creating a monument to the Pythagorean Theorem big enough to be seen from space. Twenty years later, astronomer Joseph von Littrow suggested digging trenches in geometric shapes in the Sahara Desert, to be filled with kerosene and set ablaze. Neither scheme was carried out. At the end of the nineteenth

Figure 53. The 305-meter radio dish at Arecibo Observatory in Puerto Rico represents the strengths and weaknesses of SETI. Our radio technology could detect transmitting Arecibo dishes far out into the galaxy, but this assumes alien civilizations using radio communication.

century, Jules Verne triggered a UFO scare in the United States; people read his fanciful fiction and reported seeing airships and dirigibles in the sky.[7] By the middle of the twentieth century, the US Air Force had developed slender jets, so UFO sightings took the form of sleek metal cylinders and disks. Radio SETI dominated for decades until lasers became powerful enough that researchers realized they could be used for signaling—powerful lasers can outshine a star for very brief instants of time as seen from afar.

We couldn't have done SETI a hundred years ago, and we may use quite different strategies and technologies a hundred years from now (Figure 53). SETI will only succeed if technology isn't a fleeting attribute of a civilization. As Philip Morrison pointed out in his seminal 1959 paper, "A detected signal tells us about their past and the possibility of our future."

Where Are They?

It was 1950. Enrico Fermi was visiting the lab at Los Alamos, New Mexico, where the atomic bomb had been developed. He was having lunch with three colleagues and they were discussing two seemingly unrelated items in a magazine: a spate of UFO sightings in New York and the problem of trash-can lids disappearing from city streets. They all laughed as they imagined teenagers tossing the lids past windows of apartments, making the occupants think they'd seen a UFO.

Then there was a brief pause, and Fermi said, "Where are they?"

Fermi's colleagues were used to his agile mind.[8] He was called "The Pope" by other physicists—not because he was a Catholic but because they considered him infallible on the topic of physics. Fermi had given his name to a method of estimation that allowed scientists to get the rough answer to a problem even if they had little or no data to work with. In this instance, they realized he had rapidly combined a series of suppositions: the probability that life would arise on any Earth-like planet, the vast number of planets in the Milky Way, the amount of time available for the evolution of intelligence and technology, and space exploration as a likely endeavor of an advanced civilization. When he posed the question "Where are they?" he was saying we should be surprised that the galaxy isn't littered with star voyagers. It's called "the Fermi question," and it's as well-posed today as it was in 1950.

Going a little further with this logic, we can argue that since humans got the ability to travel and communicate in space very recently, any civilization we encounter is likely to be more advanced than we are— unless we're the first to reach this level of development.

The Fermi question is provocative because it takes the failure of SETI and turns it into a poignant silence. Implicit in the question is the fact that all claims of UFOs as aliens visiting us have been without foundation. A vocal and persistent minority of the public claim sightings, encounters, or even abductions, but the scientific community believes these claims are baseless. I have spoken often enough in public to be accustomed to the UFO questions that tend to follow any astronomy talk. The most amusing (and exasperating to a scientist) involve diffuse government conspiracies and aliens kept "on ice" in Area 51. Anyone who reads the news should be skeptical that the US Government could keep such a huge discovery a secret. Carl Sagan put it best: Extraordinary claims require extraordinary evidence.

The disconnect between the expectation that spacefaring aliens exist and the lack of evidence for them has been called a paradox. That's a misuse of the word *paradox*, which is defined as any statement that contradicts itself. There are many ways spacefaring aliens can exist without our knowing about it.[9] Absence of evidence is not evidence of absence.

Let's look at some plausible answers to Fermi—which are also ways to account for SETI's Great Silence.

The first and most basic explanation is that we're alone. There are several variations of this explanation. In one, biology is a fluke and the vast numbers of potential sites for life are actually uninhabited. In another, evolution toward large brains and intelligence is highly contingent, so life almost always stays microbial. This contingency may also be astronomical. The Rare Earth hypothesis proposes that even though terrestrial planets are abundant, Earth-like planets with just the right conditions for complex life are rare. A stable long-term environment might depend on being in a part of the galaxy with the right amount of heavy elements and not too many encounters between stars, on being in a planetary system where the architecture protects against major impacts, on the planet being the right size for plate tectonics and being in a stable orbit with axial tilt stabilized by a large moon.[10] In a third variation, the development of technology and space exploration is a very unlikely outcome as species evolve. Each of these options corresponds to setting one term in the Drake equation to a very low value. In all these possibilities, we're the only intelligent, communicative civilization in the galaxy, so $N = 1$. There's no one to talk to.

The second possibility is that we're isolated. Perhaps technological civilizations do exist and some of them are plying the galaxy or sending and receiving electromagnetic signals. But if such civilizations are rare, we might be unaware of their existence. The Milky Way is 100,000

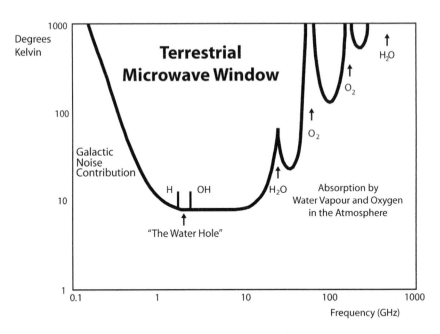

Figure 54. SETI uses the "water hole" between 1 and 10 GHz for listening and transmission because it's a cosmically quiet part of the electromagnetic spectrum. Selection of this particular frequency range assumes that alien civilizations would use a similar logic.

light years across, so if there are only ten civilizations actively exploring at any given time, the average distance between them is 10,000 light years. Texting with a 20,000-year pause between replies makes for a very stilted conversation. The information being received is old news and the sending civilization may not even exist by the time the message or probe is received. So space might be littered with the runes of dead civilizations. However, the assumption that they will use radio waves might be wrong (Figure 54). Isolation applies to time as well as space. As we've been learning, interstellar travel is expensive and difficult, so large-scale colonization may be beyond the capabilities of all but a few species around the galaxy.

The third explanation is that the search isn't good enough. Jill Tarter, the SETI pioneer who was a model for the Ellie Arroway character in *Contact*, has talked about the needle-in-a-haystack issue. The SETI haystack has nine dimensions: three of space and two each of polarization, intensity, modulation, frequency, and time. Tarter compared SETI to scooping a bucket of water from an ocean in the hope of catching a fish. That's going to change as the Allen Telescope Array reaches its full potential. An irony of working in a field where detection capability improves exponentially is that each new search is better than the sum of all the searches that preceded it.[11] The Allen Array will survey a million stars for artificial signals over a frequency range from 1 to 10 gigahertz.

Tarter refuses to be discouraged by the Great Silence. Rather, she thinks the search is only just beginning to get interesting. Our most powerful "send" capability is the radar transmitter at the Arecibo 305-meter radio dish in central Puerto Rico.[12] The Allen Array will be able to detect analogous technology beamed from a planet around any star out to a thousand light years. Optical SETI's ability to detect pulsed lasers is also getting better. In twenty years, SETI could detect the equivalent of our most powerful radio and optical transmitters if they are beaming at us from planets around 100 million stars in the galaxy. At that point, surely, continuing Great Silence means we're alone.

Or perhaps the search is ill conceived. We might be looking in the wrong part of the cosmic haystack. Denizens of other planets may be using forms of data modulation and compression that look to us like pure noise. Radio transmitters and laser might be fleeting technologies, so the window of time when they get used by a civilization is small. Seth Shostak, a senior astronomer at the SETI Institute, jokes about people who say he should wait until he has more powerful technology, noting that Queen Isabella of Spain didn't tell Christopher Columbus to wait until he had a jumbo jet to discover America.

Another possibility is that all other life is already dead. When it comes to putative alien physiology and psychology, answers to the Fermi question proliferate like weeds. As we see ominously from our own history, galloping technology might render a civilization unstable. If the factor L in the Drake equation is on average no more than a few centuries or a millennium, SETI is likely to fail. It doesn't matter whether the civilizations self-destruct or degrade to a preindustrial state—the effect on the search is the same. Silence.

It's also possible that extraterrestrial life is unrecognizable to us. Aliens in films and on TV are amusing because they're usually thinly veiled versions of us: bipedal vertebrates with extra appendages or bad skin. Occasionally, they're amorphous and weird. But life elsewhere might be organized quite differently at the level of the organism. It might communicate at speeds that are too slow or too rapid for us to recognize. It might not be culturally inclined to communicate at all, or it might be disinterested in space travel even if it had the suitable technology. It might have passed from a biological to a post-biological or computational state. We can try to think outside the box, but anthropocentrism permeates all discussions of advanced life beyond Earth.

The Great Filter

Does the lack of evidence for extraterrestrial technology say anything about us and our future?

Yes, especially if spacefaring civilizations are very rare, as opposed to just hard to detect or recognize. In 1998, economics professor Robin Hanson presented an idea called "the Great Filter." If any of the Drake equation factors is very low, it will act as a filter to choke off evolution toward life venturing beyond its planet of origin. A filter can lie behind us (in our past), or ahead of us (in our future).[13] In the past, the filter could be the transition from single- to multicelled organisms, or the steps required to develop a brain, or the instability of a technological species. When humans harnessed the power of the atomic nucleus in the middle of the twentieth century, it was in the unstable and destructive form of a bomb. For a decade, we teetered on the edge of nuclear holocaust. Regarding the future, this argument leads to the counterintuitive and unnerving conclusion that the easier it is for life to get to our stage of development, the bleaker our future chances for survival probably are.

Assuming that a Great Filter winnowed down billions of germination sites for life to zero observable extraterrestrial civilizations, it's crucial to consider where the filter lies. If the filter is in our past, it means there is an extremely unlikely step in the progression from Earth-like planet to one that hosts a civilization with our level of technology. That step might even be the formation of life from simple chemicals. Whatever the filter is, if it's behind us, we can explain the lack of observable aliens. Technological civilizations are intrinsically rare, so searches for them will fail.

On the other hand, if the Great Filter is in our future, it's very unlikely that a civilization at our stage of development progresses to the large-scale colonization of space. One plausible scenario is that technology is the culprit because it includes the capability for self-destruction. Nick Bostrom, director of the Future of Humanity Institute at the University of Oxford, has done scholarly work on catastrophes. His partial list of existential threats faced by humanity includes nuclear holocaust, genetically engineered superbugs, environmental disasters, asteroid impacts, terrorism, advanced and destructive artificial intelligence, uncontrollable nanotechnology, catastrophic high-energy physics experiments, and a totalitarian regime with advanced surveillance and mind-control technologies.

Regarding existential threats that might act as a filter in our future, Bostrom makes another point. The requirement is not that it has a significant probability of destroying humanity. Rather, it must be able to plausibly destroy *any* advanced civilization. Asteroid strikes and supervolcanoes don't qualify because they're random events that some civilizations will survive and others won't experience because their planet and solar system are different from ours. The technological innovations

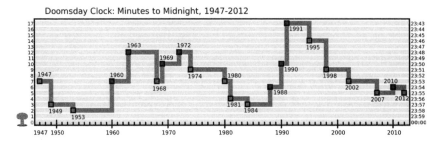

Figure 55. In the short history of the "nuclear age," we have come close to a holocaust several times. The Doomsday Clock tracks our proximity to Armageddon. Civilizations may become unstable and destroy themselves. This issue impacts the prospect of companionship and contemporaneous communication in space.

that drive the argument and act more effectively as filters are those that almost all civilizations eventually discover, where their discovery almost universally leads to disaster (Figure 55).

Bostrom has said: "I hope that our Mars probes will discover nothing. It would be good news if we find Mars to be completely sterile. Dead rocks and lifeless sands would lift my spirits."[14] Why would he be so grumpy about one of our best pieces of technology?

If life is discovered on Mars or any other place in the Solar System, it suggests that the emergence of life is not an improbable event (and it doesn't matter whether the life found is ancient or current). If biology emerged twice independently in our backyard, then surely there are many biological experiments in the galaxy. The same logic will apply if we one day find that a significant number of habitable exoplanets have had their atmospheres altered by microbial life. Either discovery would imply that the Great Filter is less likely to occur in the early history of planets and is more likely ahead of us. In other words, dead rocks and lifeless planets will be good news since they'd tell us we'd survived the tough part of our evolution.

This framing of the argument is a simplification. There may be more than one Great Filter. We might have cleared one, only to be faced in the future with another. Also, we should be wary of positing that life elsewhere has to follow the path traveled by life on Earth or that other civilizations progress with the single-minded purposefulness of humans. Let's allow Nick Bostrom to have the last word. For someone who dwells on apocalypse and hopes that the search for life in the universe fails, he's strangely optimistic:

> If the Great Filter is in our past . . . we may have a significant chance . . . of one day growing into something almost unimaginably greater than we are today. In this scenario, the history of humankind to date is a mere instant compared to the eons of history that still lie before us. All the triumphs and tribulations of the millions of people who have walked the Earth since the ancient civilization of Mesopotamia would be like mere birth pangs in the delivery process of a kind of life that hasn't really yet begun.

14

A Universe Made for Us

Our Far Future

If we make it through our troubled adolescence as a species, what lies in store? We're curious and creative but also prone to tribalism and needless competition. I've sketched a scenario for the next century, when we establish homesteads on the Moon and Mars, project our tourism and commerce off-Earth, and get used to traveling throughout the Solar System.

By overcoming our self-destructive tendencies, we might achieve the normal evolutionary lifespan of a mammal species, a million years or more. To see how hard it is to project ourselves forward that far, let's play at "futurology." Compressing time in orders of magnitude, we first look backward. Roughly ten years ago, there was no Internet. Roughly a hundred years ago, there was no mass transit and most people lived and died close to where they were born. Going back a thousand years, there was no medicine and life was short and brutal. Ten thousand years ago, agriculture would soon be invented but most humans were nomadic hunters and gatherers. Approximately a hundred thousand years ago, we hadn't learned how to use tools or harness fire. A million years ago marks our emergence as a species. Leaping back in factors of ten quickly finds us in an unfamiliar and primitive state (Figure 56).

Now play the game forward. It's fairly safe to predict that in ten years we'll have sophisticated genetic engineering and a growing commercial space industry. In a century, we should have routine travel within the Solar System, robots doing our bidding, and artificial intelligence that rivals human capabilities. It's very difficult to predict a thousand years hence, but I'm going to go out on a limb and assume that rapid technological progress will continue such that some of us will be heading for nearby stars. Ten thousand years from now, as far ahead of us as early civilizations lie behind us, the crystal ball is cloudy. A hundred thousand years and onward, it's anyone's guess. To venture further seems impos-

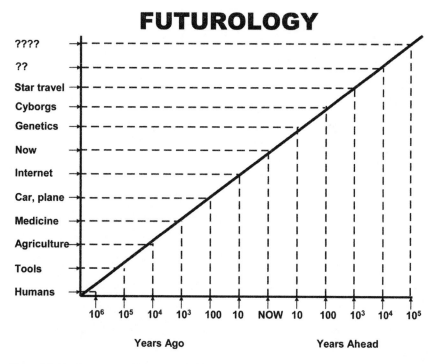

Figure 56. The human past becomes primitive viewing history in orders of magnitude of time. Landmarks in the past are labeled, along with speculations about the next thousand years. If humans persist for millions of years, our capabilities could be profoundly more advanced than they are now, and very difficult to predict.

sible. In the anthology *Year Million*, economists, science fiction writers, computer scientists, and physicists speculate freely, in tones that range from giddy to glum.[1] The exercise is made more difficult by our position on the cusp of exponential technological change.[2] The far future, with all its potential for our greatness and failure, and the possibility that we might not exist at all, is as haunting as deep space.

When we finally leave the Solar System, those voyagers will represent a slender green shoot from the sturdy human tree. There's no need to break laws of physics, or even to travel at a substantial fraction of the speed of light. They will never rejoin the tribe. Once they leave home, there's no returning. The first European settlers to America knew that they would never go home; the commitment of the first star travelers will be just as absolute. They will, however, need to be kept alive for the duration of the trip.

Exploration of space beyond the Solar System is only possible if we persist as a species. It will also require innovations to extend the lives of the travelers.

The pig is named 78-6. She's ruddy pink and weighs 120 pounds, and her beating heart is exposed in the operating theater. The surgeon cuts the aorta, watches the EKG flatline, and connects external tubes to replace the pig's blood with a chilled saline cocktail. With her vital organs preserved, 78-6 isn't quite dead. She's in a state of cryogenic suspension or suspended animation at Massachusetts General Hospital in Boston. This surgeon has suspended 200 pigs for one or two hours each, and they all survived as long as they were given optimal treatment. A few hours later, 78-6 will wake up in a recovery room with classical music on the radio and a healthy pig in an adjacent stall for company. Postmortem exams on other pigs showed no cognitive damage from the procedure.[3]

Research in suspended animation is in its early stages. Pigs are critical test subjects because their physiology is close to that of humans. Also at Mass General, mice have had their metabolisms slowed by factors of ten or twenty. In other research labs, dogs have been revived after hours of being clinically dead. In a few cases, humans have survived being subjected to extreme hypothermia and near death for several weeks.[4]

But we're taking the long view now, so let's assume that we eventually master the art of suspended animation. This finesses the obstacle of long travel times—these starship Rip Van Winkles would wake up and continue with their new lives, oblivious to the centuries it took to get to their destination. Suspended animation would create an irrevocable rift between the voyagers and the Earth. Everyone they knew and loved, and their descendants, would have lived and died while they silently sailed through the void.

Let's also assume that human cloning will one day be perfected. Since the pioneering experiment that gave birth to Dolly the sheep in 1996, cloning has been performed uneventfully on rabbits, goats, cows, cats, and fifteen other species.[5] Primate reproductive biology appears to be more complex, but it's only a matter of a few years before humans are cloned. Cloning is ethically fraught, but it would provide a way for us to propagate in the vastness of space. Instead of one set of colonists, chosen to be a minimum viable population with a good genetic mix, there would be a suite of colonies, each composed of the same set of cloned individuals. Each cloned colony would disperse to a different destination. Each would grapple with a different environment. Despite identical DNA, the evolutionary paths of the colonies would diverge. Taken together, they would play out natural selection on a new cosmic stage.

How then will we head for the stars?

Conceptually, there are four approaches: The travelers live and die on the spaceship, they travel in suspended animation, they're carried as embryos or single cells, or they're transported digitally at the speed of light. These four scenarios are ordered in increasing level of technical sophistication but decreasing order of resources required for the trip.

We've seen that vast Gerard O'Neill pinwheels loaded with thousands of passengers are ruinously expensive, and teleportation is far beyond current and projected technology. Suspended animation is promising, and it need not be for the whole trip. Subsets of the crew could be revived periodically to monitor life-support systems and carry out routine maintenance. Embryo transport may also be possible one day.[6]

Echoing Arthur C. Clarke, we've imagined the emotional impact of the cry of the first off-Earth infant. But what if that baby is animated after a journey of millennia and tens of light years? It would have traveled as a frozen zygote, the earliest developmental stage of an embryo. It would then have been brought to term in an artificial womb, and reared to self-sufficiency by robotic nannies, all to be part of a new human colony. The starship would also carry frozen cells of useful livestock and crops, serving as a miniaturized Noah's ark.

Living in the Multiverse

We leave a tiny footprint in the universe. The sum of all our industry and striving is a spherical ripple moving out into the void. We've had powerful radio and TV transmitters operating for fifty years, and in principle that expanding sphere of radiation has swept over thousands of habitable worlds. In practice, all the pop-culture messages carried by these waves are diluted to a level below the hiss of the cosmic background radiation before they exit the Solar System. The Pioneer and Voyager spacecraft carry information about our civilization, and they're our first artifacts to reach interstellar space, but it will be hundreds of thousands of years before they reach another star.

Other creatures to have left their planet are likely to be considerably more advanced than us. What would that imply?

Since it's impossible to anticipate the function and form of an alien species, the simplest way to categorize hypothetical civilizations is by their energy use. This was first done by the Russian Nikolai Kardashev. Kardashev studied astronomy while both of his parents were in Stalin's slave-labor camps in the 1950s. He heard about Frank Drake's Project Ozma, which inspired him to write his influential paper "Transmission of Information by Extraterrestrial Civilizations."[7] In it, he defined three levels based on the amount of power available to a civilization. Type I civilizations utilize all the solar energy arriving at their planet's surface, about 10^{17} watts for a planet like the Earth and a star like the Sun. The next tier is 10 billion times higher—a Type II civilization harnesses all the energy from its star, about 10^{27} watts. A Type III civilization is 10 billion times hungrier, consuming energy at the phenomenal rate of 10^{37} watts, the luminosity of a galaxy like the Milky Way. Beyond the original Kardashev scale is Type IV: masters of the universe (Figure 57).

Kardashev created his scale to categorize technologically mature civilizations. We're so feeble that we don't even make it onto the scale. Stuck getting energy from dead plants, our vaunted civilization runs on a measly 0.001 percent of the energy that arrives gratis from the Sun. Theoretical physicist Michio Kaku has noted that with our energy consumption growing at 3 percent per year, we'll rise to Type I status in a few centuries, Type II status in a few millennia, and, if we make it that long, Type III status in a million years.[8]

Type I civilizations will elude detection, giving off extra waste heat but not enough to be detected from many light years away. Civilizations that harness most of their star's energy might be detectable because they would have to build something like a Dyson sphere.[9] Freeman Dyson published this thought experiment in 1960, based on a 1937 Olaf Stapledon science fiction novel. The idealized concept of a hollow sphere around a star is physically unstable (in Larry Niven's *Ringworld* series of science fiction novels, this instability causes a collapse of the civilization), but a civilization might build a swarm of orbiting satellites to envelop the star and capture most of its energy. The visible light is captured and reradiated as infrared emission, so Dyson spheres are detect-

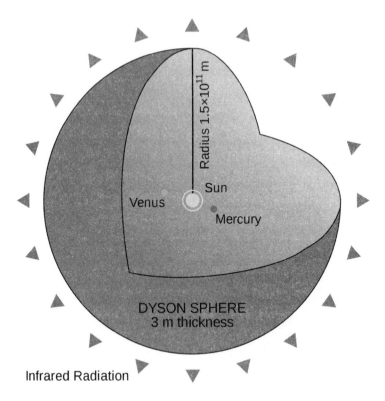

Figure 57. A Dyson sphere is the theoretical concept of an energy collection system that can harvest all the radiation from a star. Nikolai Kardashev imagined a scale of increasing energy usage as a civilization matured—from using the energy falling on a planet (I) to using the energy of a star (II), a galaxy (III), and the universe (IV). A Dyson sphere is the technology of a Type II civilization while we are currently less than a Type I civilization.

able as excess infrared emission from an otherwise normal star. Several SETI projects look for anomalous infrared radiation. Researchers at Fermilab, near Chicago, sifted 250,000 stars down to seventeen candidates, of which four were declared "amusing but still questionable."[10]

The existence of Dyson spheres allows for passive SETI, where no intention to communicate is needed. The premise is that any highly advanced civilization will leave a much larger footprint than we will. Type II or later civilizations may employ technologies that we're tinkering with or can barely imagine. They might orchestrate stellar cataclysms or use propulsion by antimatter. They might manipulate space-time to create wormholes or baby universes, and communicate via gravity waves. We can look for artifacts as well as messages. Extrapolation is addictive, so some scientists have proposed adding the category of a Type IV civilization that controls space-time well enough to affect the entire universe.

Why stop at one universe?

Modern cosmology involves the idea of quantum genesis—tracing back the cosmic expansion projects to an origin in a singularity where the space that now contains 100 billion galaxies was smaller than an atom. The inflationary scenario is an adjustment of the standard big bang to include an extremely early phase of exponential expansion. The idea was developed to explain why the universe now is very smooth and geometrically flat. Inflation has tentative support from the nature of small temperature variations in the cosmic background radiation. If inflation is correct, the universe began as a quantum fluctuation. The precursor state would have been an ensemble of quantum fluctuations, perhaps infinite in number, each with randomly different initial physical conditions. Some of them inflated into large space-times like our own. Others were stillborn. This process can be timeless and eternal (Figure 58). The laws of nature in these parallel universes would differ

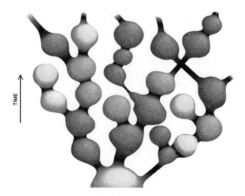

Figure 58. In chaotic inflation, the precursor state was an endless series of space-time quantum fluctuations. Some of these fluctuations might inflate into macroscopic universes, while others would not. This is the multiverse concept.

from the laws with which we're familiar.[11] This, in a nutshell, is the multiverse.

The multiverse is connected to another issue that has been perplexing physicists for several decades: fine-tuning.[12] Albert Einstein fervently believed that the laws of physics, when they were fully understood, would be inevitable, elegant, and self-contained. This quality, called naturalness, has been a touchstone in theories of nature ever since. But nature isn't cooperating. The "standard model of particle physics" precisely explains the interactions of fundamental particles, but the model is governed by more than two dozen parameters, so it's not elegant or simple and the parameters don't emerge naturally from an underlying theory. Some quantities—like the mass of the Higgs particle and the value of the dark energy that controls cosmic acceleration—are much lower than physicists expected. They're dismayed that laws of nature seem to be an arbitrary and messy outcome of random fluctuations in the fabric of space-time.[13]

A controversial argument deriving from fine-tuning is the fact that the forces of nature and the attributes of the universe appear to take values required for carbon-based life to exist. If the electromagnetic force was stronger or weaker, stable atoms could not form. If the strong nuclear force was stronger or weaker, carbon couldn't be created in stars. If the gravitational constant was stronger, stars would be very short-lived; if it was weaker, stars wouldn't shine or make the heavy elements. The universe also has very low entropy or disorder, which may be responsible for time's forward sense or "arrow." In addition, the cosmic values of dark matter and dark energy neither prohibit structures from forming nor cause a collapse too soon for life to be able to form. The point is that while the universe would be physically sensible if any of these quantities took different values, it wouldn't be a universe with life containing life as we know it.

It's unremarkable that the properties of the universe are compatible with our existence. But controversy arises when this anthropic line of reasoning is strengthened to say that the universe *must* have those particular properties that allow life to develop at some point.

Inflation and (as yet unproven) string theory could provide a physical basis for a vast ensemble of parallel universes with randomly different properties. Our universe is "special" only in the sense that it contains biology.

Debate rages over whether the multiverse concept is real science or solipsism. Meanwhile, Alan Guth, the MIT physicist who developed inflation theory in the 1980s, suggests it might provide yet another explanation for the absence of aliens. Assuming plausible probabilities for the multiverse ensemble, young universes vastly outnumber the old ones.[14] Averaged over all universes, universes with civilizations will almost always have just one, the first to develop: us.

Singularity and Simulation

The conjecture that civilizations more advanced than ours will use an ever-increasing amount of energy is very retro, very twentieth century. In fact, the rate of growth of world energy consumption peaked at 5 percent per year in the 1970s, and it has fallen to less than 3 percent per year now. World energy appetites are likely to rise at even lower rates in the future as fossil fuels become depleted, population growth slows, and the cost of energy drives industry to greater efficiency. We may never reach the usage of a Type I civilization, let alone the energy-guzzling heights of the Type II and Type III civilizations.

Rather than become grandiose and bloated, an advanced civilization might slim down. Earth's population and resource use may stabilize within a few decades. But two exponential trends will likely continue unabated: shrinking physical devices through nanotechnology and growing computational power and information storage. Let's look at each of these in turn.

Physicist John Barrow has created an alternative to the Kardashev scale for classifying civilizations based on their ability to manipulate matter. The scale progresses from manipulating human-scale objects to manipulating molecules to make new materials, then to manipulating atoms and creating new artificial life forms. The third level is almost within our grasp. Beyond that is the ability to manipulate elementary particles and make entirely novel forms of matter, culminating in the manipulation of the basic structure of space-time. The self-replicating von Neumann space probes that we encountered earlier are incredibly cumbersome when compared with what could be done with self-replicating nanomachines.[15]

The frontier of controlling matter at the level of fundamental particles has been explored by artificial intelligence (AI) researcher Hugo de Garis. He writes: "The hyperintelligences that are billions of years older than we are in our universe have probably 'downgraded' themselves to achieve hugely greater performance levels. Whole civilizations may be living inside volumes the size of nucleons or smaller." Alien artifacts may be built into the architecture of matter, leading to a new paradigm for SETI: "Once one starts 'seeing' intelligence in elementary particles, it changes the way one looks at them, and the way one interprets the laws of nature, and the interpretation of quantum mechanics, etc. It's a real paradigm shift away from looking for non-human intelligence in *outer* space, to looking for it in *inner* space."[16]

Next we turn to the progress in computation. Exponential gains in processing power lead to the idea of the technological singularity. This is the time, projected to be in the middle of the twenty-first century, when civilization and human nature itself are fundamentally transformed. One variant of the singularity is when artificial intelligence surpasses human intelligence. Software-based synthetic minds begin to program themselves and a runaway reaction of self-improvement occurs. This event was foreshadowed by John von Neumann and Alan Turing in the 1950s. Turing wrote that ". . . at some stage therefore we should have to expect the machines to take control . . . ," and von Neumann described ". . . an ever-accelerating progress and changes in the mode of human life, which gives the appearance of approaching some essential singularity in the history of the race beyond which human affairs, as we know them, could not continue."[17]

A dystopian version of this event permeates the popular culture, from science fiction novels to movies such as *Blade Runner* and *The Terminator*. Doctor Frankenstein is destroyed by the powerful monster he creates—it's a venerable morality tale. The possibility that advanced civilizations might be aggressive is the reason Stephen Hawking argues that we shouldn't try to communicate or reveal our presence. The most chilling example of this scenario is seen in Fred Saberhagen's Berserker series of novels, where self-replicating doomsday machines are out there watching, ready to destroy life on a planet just as it begins to acquire advanced technology.

Another variant of the singularity takes current efforts to fight disease and projects them into radical life extension, where technology helps us overcome all our mental and physical limitations. Ray Kurzweil has been the most eloquent proponent of this future. He's a founder of the Singularity University, where the tech world's movers and shakers pay tens of thousands of dollars for short courses on the cutting edge in AI and nanotechnology. Critics have mocked the idea of the singularity as the "rapture of the nerds," and they've noted that only the wealthy will benefit from radical-life-extension technology.

The goal of researchers like Kurzweil is simple: immortality. He thinks that medical nanotechnology will conquer disease, aging, and death. To become postbiological, we'll need the means to reverse engineer the brain of any human and reproduce it in silicon. The term for this is *mind uploading*.

Hans Moravec first outlined the simulation argument in the late 1990s. He projected advances in computation to superhuman capabilities in a few decades. With the premise that the wet electrochemi-

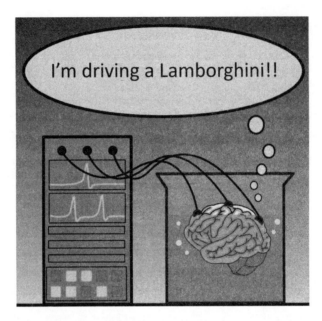

Figure 59. There is a long philosophical tradition involving the thesis that reality is an illusion. A modern version of this idea posits that humans are all simulated entities of an advanced alien civilization.

cal network of the brain, including consciousness, can be replicated computationally, we're within striking distance of a computation that's equivalent to the history of all the thoughts every human has ever had. If one computer can simulate the thoughts of a person, a suitably powerful computer could simulate the entirety of human consciousness. And if we can do that, it would be trivial for any more advanced civilization. Following Nick Bostrom, let's call this capability "technological maturity." It's a modern version of the concept in philosophy known as "the brain in a vat" (Figure 59). This scenario is familiar from movies of *The Matrix* franchise, where humans are all simulated by a superior civilization, but there are hints of that fact and some learn how to control the simulation. The movies are provocative but somewhat illogical: A simulation that sophisticated would not have "glitches" that revealed to the simulated entities that they were in a simulation.

Philosopher Nick Bostrom has explored the consequences of replicating human consciousness in a computer. He's formalized the scenario into a simulation hypothesis. Based on formal logic, you must accept one or more of the following propositions: (1) Humans will go extinct or self-destruct before we become technologically mature and able to carry out such simulations. (2) No other civilization that can create such simulations does so. (3) We live in a simulation.[18]

Most people react with shock and revulsion to this prospect. After all, we each carry with us our own crisp sense of identity and reality. But the argument deserves to be taken seriously. We might hope to reject the first proposition, since it implies a gloomy outcome for humanity—there's no reason to believe we'd be luckier than any other civilization in achieving maturity. The second proposition seems unlikely, since it would require a commonality of purpose among disparate civilizations. If any civilizations are creating these simulations, it's so easy to create a vast number of simulated entities like us that they would outnumber flesh-and-blood creatures. The principle of mediocrity says we're more likely to be simulated than real. Bostrom admits it's difficult to assign probabilities to the correctness of the three propositions, but he argues that there's no reason to assign low probability to the third.

Philosophers have even written papers about how you should live if you think you are in a simulation. The simulation designers wouldn't give you a hint of your predicament unless they wanted to, but they might mix sentient with nonsentient creations, leaving you to figure out who the zombies are. If we live in a simulation, then space travel isn't as adventurous as we imagine; it's like playing a video game within a video game. And there's no reason to believe the simulators are not themselves simulated, leading to infinite regression—a problem philosophers and logicians still haven't figured out.

If these bizarre ideas have any validity at all, what do they say about our urge for exploration and for leaving the Earth? If reality is a chimera and we are simulated, then space travel is part of the simulation. It's no more difficult or meaningful than riding a bicycle. Even if we reject this possibility, we can still admit that advanced civilizations may use their powers to create fantastical computations or simulated entities. They could have rich inner lives. They could be purely contemplative. But they'd be entirely bounded by their own capabilities. It would be a hermetic, self-defined existence.

Like all sentient beings, we have a choice. We can turn inward or turn outward. So far, humans have chosen to explore and to venture into the unknown. It's not the easy choice and sometimes it carries great risk. But it's a bracing way to live our finite lives.

Imagination and Exploration

We stand at the edge of a vast cosmic shore. We've dipped our toes in the water and found it bracing but inviting. Time to jump in.

Imagination is one of a human being's most singular gifts, and we've used it to create fully realized worlds in art, music, fiction, and poetry. Science is not just a dry collection of facts and theory—it's driven by imagination. When Newton imagined a cannon on top of a high mountain firing a projectile that would fall at just the same rate that the Earth curved below it, he was imagining space travel more than two centuries before we would develop the technology to leave the Earth's gravity. Science fiction writers and space artists have long dreamed of other worlds, and the exoplanets being discovered in droves are exotic enough to measure up to their visions.

Let's not forget how we got to this point. Animals roam, they seek food, and they expand their range. But only humans have the urge to explore for its own sake. We left Africa and by 10,000 years ago we had forged better lives for ourselves by taming plants and animals and settling into fixed communities. But the curiosity remained. So we harnessed the wind to sail the oceans. Then we loosened the bonds of gravity by using chemical fuels to fly in planes and soar in rockets. The challenges of space exploration have already led us to develop better engines, faster computers, and smarter materials. In the future, space exploration will drive us to develop efficient fuels, miniaturized control systems, and sophisticated medical diagnostics.

Space travel is urgent and it is real. We have the technology and the means to live and work in space, gain a permanent toehold off-Earth, and explore the Solar System and beyond. No laws of physics stand in our way. If we commit to space travel, it will force us to cooperate as a species, since some of the problems are too hard for any one country to solve. The effort can ennoble us.

Space travel, however, will never be our top priority. There are poor people to feed, diseases to cure, wars to resolve, and a bruised planet to heal. Yet venturing beyond the Earth challenges our ingenuity in ways that can benefit everyday life. Learning about other worlds informs us how to be better caretakers of this one. These activities can let us be more than a footnote in the history of the Milky Way. Exploration is built into our DNA; we should not resist. Everyone deserves the chance at least once in their lives to be liberated from the heft of the body's gristle and to view the jewels of the cosmos set against the black velvet of night.

Notes

1: DREAMING OF BEYOND

1. The Genographic Project, sponsored by the National Geographic Society, has used DNA from nearly a million people to map out human migration. See https://genographic.nationalgeographic.com/about/.

2. Darwin speculated about a tree of life and a last common ancestor based on morphological similarities between species. The size and shape of an organism can be misleading, however, and bacteria are all the same shape, so modern phylogeny uses measures of overlap in the base pair sequences of DNA or RNA. It's hard to reconstruct linear time using genetic distance, and gene transfer and convergent evolution can cause confusion. When there are many species being compared, more than one tree may fit the data.

3. "Resolving the Paradox of Sex and Recombination" by S. P. Otto and T. Lenormand 2002. *Nature Reviews Genetics*, vol. 78, pp. 737–56.

4. The Geno 2.0 testing kit costs $200 online. After someone sends in a cheek swab, individual results are returned showing the broad pattern of ancestry and degree of genetic overlap with different native populations. The research results are in "The Genographic Project Public Participation Mitochondrial DNA Database" by D. M. Behar et al. 2007. *PLoS Genetics*, vol. 3, no. 6, p. e104.

5. "The Arrival of Humans in Australia" by P. Hiscock 2012. *Agora*, vol. 47, no. 2, pp. 19–22.

6. "How Babies Think" by A. Gopnik 2010. *Scientific American*, July, pp. 76–81. See also *The Scientist in the Crib: Minds, Brains, and How Children Learn* by A. Gopnik, A. N. Meltzoff, and P. K. Kuhl 1999. New York: William Morrow and Company.

7. The characterization of some DNA as junk is likely to reflect our ignorance in recognizing the way DNA triggers genes into expression in the organism. In 2008, a study led by James Noonan of Yale University found that a small region of non-coding or "junk" DNA was responsible for developments in the ankle, foot, thumb, and wrist that were key evolutionary changes, allowing us to walk upright and use tools.

8. "Population Migration and the Variation of Dopamine D_4 Receptor (DRD_4) Allele Frequencies Around the Globe" by C. Chen, M. Burton, E. Greenberger, and J. Dmitrieva 1999. *Evolution and Human Behavior*, vol. 20, no. 5, pp. 309–24.

9. "Cognitive and Emotional Processing in High Novelty Seeking Associated with the L-DRD_4 Genotype" by P. Roussos, S. G. Giakoumaki, and P. Bitsios 2009. *Neuropsychologia*, vol. 47, no. 7, pp. 1654–59.

10. "Learning about the Mind from Evidence: Children's Development of Intuitive Theories of Perception and Personality" by A. N. Meltzoff and A. Gopnik 2013, in *Understanding Other Minds: Perspectives from Developmental Social Neuroscience*, ed. by S. Baron-Cohen, H. Tager-Flusberg, and M. Lombardo 2013. Oxford: Oxford University Press, pp. 19–34.

11. "Causality and Imagination" by C. M. Walker and A. Gopnik, in *The Development of the Imagination*, ed. by M. Taylor 2011. Oxford: Oxford University Press. See also "Mental Models and Human Reasoning" by P. N. Johnson-Laird 2010. *Proceedings of the National Academy of Sciences*, doi/10.1073/pnas.1012933107.

12. *A Brief History of the Mind* by William Calvin 2004. Oxford: Oxford University Press.

13. "The Cognitive Niche: Coevolution of Intelligence, Sociality, and Language" by S. Pinker 2010. *Proceedings of the National Academy of Sciences*, doi/10.1073/pnas.0914630107.

14. "The Human Socio-cognitive Niche and Its Evolutionary Origins" by A. Whiten and D. Erdal 2012. *Philosophical Transactions of the Royal Society B* [Biological Sciences], vol. 367, pp. 2119–29.

15. "Plurality of Worlds" by F. Bertola, in *First Steps in the Origin of Life in the Universe*, ed. by J. Chela-Flores et al. 2001. Dordrecht: Kluwer Academic Publishers, pp. 401–7.

16. "Anaxagoras and the Atomists" by C. C. W. Taylor, in *From the Beginning to Plato: Routledge History of Philosophy, Vol. 1*, ed. by C. C. W. Taylor 1997. New York: Routledge, pp. 208–43. See also "The Postulates of Anaxagoras" by D. Graham 1994. *Apeiron*, vol. 27, pp. 77–121.

17. *On the Nature of Things* by Lucretius Carus, trans. by F. O. Copley 1977. New York: W. W. Norton.

18. For an overview of the conceptual leaps made by a small number of bold thinkers 2,500 years ago, see *The Presocratic Philosophers* by J. Barnes 1996. New York: Routledge.

19. It would be inappropriate to infer a modern cosmological context from the plural-

ism of world religions. For example, the "many worlds" in Buddhist texts are part of a geocentric cosmology with Mount Mehru as the central feature, and no distances are assigned to these remote regions, which are constantly coming into and out of existence.

20. "The True, the False, and the Truly False: Lucian's Philosophical Science Fiction" by R. A. Swanson 1976. *Science Fiction Studies*, vol. 3, no. 3, pp. 227–39.

2: ROCKETS AND BOMBS

1. Humans are built for throwing a projectile forward rather than upward, given our origins as hunters. The fastest pitch in baseball is around 105 mph. If directed upward, that would reach a height of 70 meters. The British javelin thrower Roald Bradstock holds many official and unofficial world records for throwing items as diverse as a dead fish and a kitchen sink. His horizontal record is 130 meters for a cricket ball and 160 meters for a golf ball; the latter would be equivalent to 80 meters if thrown vertically. If you want to try your hand at a vertical toss, you can use your smart phone, after you install an app called "Send Me to Heaven."

2. "The History of Rocketry, Chapter 1" by C. Lethbridge, hosted by the History Office at NASA's Marshall Space Flight Center, online at http://history.msfc.nasa .gov/rocketry/.

3. *Throwing Fire: Projectile Technology Through History* by A. W. Crosby 2002. Cambridge: Cambridge University Press, pp. 100–103.

4. Gunpowder has a rich and complex history. It's characterized as a "low" explosive that deflagrates, as opposed to a "high" explosive like TNT that detonates. Gunpowder was invented by Chinese alchemists who were trying to create a potion for eternal life. Saltpeter or potassium nitrate had been used by the Chinese for medicine since the first century AD, and in gunpowder it acts as an oxidizer. Sulfur and charcoal act as fuel.

5. *Science and Civilization in China: Vol. 3, Mathematics and the Sciences of the Heavens and the Earth* by J. Needham 1986. Taipei: Cave Books Ltd., p. 104.

6. The "bible" of rocketry for more than two hundred years, from the mid-seventeenth to the mid-nineteenth century, was *The Great Art of Artillery* by Kazimierz Siemienowicz. It contained an array of designs, including multistage rockets and rockets with stabilizing delta wings.

7. It was a brilliant realization of Isaac Newton that an object falling due to gravity is undergoing the same motion as an object in Earth orbit. He had formulated gravity as an inverse square law and the Moon is sixty times farther from the center of the Earth than someone standing on the Earth's surface, so the gravitational acceleration will be 3,600 times smaller for the Moon than for a cannonball. The deviations of the Moon and the cannonball in their trajectories are related in exactly the way expected for an inverse square force law.

8. It's called the "ideal" rocket equation because it only holds true for reaction-engine vehicles where the exhaust velocity is constant or can be effectively averaged. No aerodynamic or gravitational effects are included, and it only applies under the assumption that propellant is discharged and the delta-v applied instantly. For a multistage rocket, the equation applies separately for each stage.

9. Tsiolkovsky is rightly famous for his pivotal role in the theory of spaceflight. However, his equation was actually first derived and published in a pamphlet more than a century earlier by British mathematician William Moore, working at the Royal Military Academy in Woolwich. See *A Treatise on the Motion of Rockets* by W. Moore 1813. London: G. and S. Robinson.

10. *The Red Rockets' Glare: Spaceflight and the Soviet Imagination, 1857–1957* by A. A. Siddiqi 2010. Cambridge: Cambridge University Press, pp. 62–69.

11. *Investigations of Outer Space by Rocket Devices* by K. Tsiolkovsky 1911, quoted in *Rockets, Missiles, and Men in Space* by W. Ley 1968. New York: Signet/Viking.

12. *The Russian Cosmists: The Esoteric Futurism of Nikolai Fedorov and His Followers* by G. M. Young 2012. New York: Oxford University Press.

13. As a young man, Oberth was a consultant on *Woman in the Moon*, the first film ever to have scenes set in outer space, directed and produced by the great Fritz Lang. Oberth built rocket models for the film and launched a rocket as a publicity stunt for the film's opening. Decades later, he was given a nod in the *Star Trek* films and TV series, which named a class of starships after him.

14. "Hermann Oberth: Father of Space Travel," online at http://www.kiosek.com/oberth/.

15. *The Autobiography of Robert Hutchings Goddard, Father of the Space Age: Early Years to 1927* by R. H. Goddard 1966. Worcester, MA: A. J. St. Onge.

16. Lindbergh and Goddard formed a lifelong alliance and friendship because they shared a dream of travel beyond the Earth. The famous aviator helped Goddard to get funding when no government agency would support him and few were taking his work seriously. He eventually received long-term support from the financier and philanthropist Daniel Guggenheim. After Goddard's death, his estate and the Guggenheim Foundation successfully sued the US Government for patent infringement. At the time, the award of $1 million was the largest settlement ever in a patent case.

17. *New York Times*, "Topics of the Times," January 13, 1920, p. 12.

18. *New York Times*, "A Correction," July 17, 1969, p. 43.

19. *Rocket Man: Robert H. Goddard and the Birth of the Space Age* by D. A. Clary 2004. New York: Hyperion, p. 110.

20. Quoted in "Rocket Man: The Life and Times of Dr. Wernher von Braun" by K. Baxter 2006. *Boss* magazine, Spring, pp. 18–21.

21. "Recollections," early experiences in rocketry as told by Wernher von Braun 1963, hosted by the History Office at NASA's Goddard Space Flight Center, online at http://history.msfc.nasa.gov/vonbraun/recollect-childhood.html.

22. Von Braun admitted in 1952 that he "fared relatively rather well under totalitarianism." This and further analysis of his ambiguous relationship to the Nazi regime and weapons of mass destruction are reviewed in "Space Superiority: Wernher von Braun's Campaign for a Nuclear-Armed Space Station, 1946–1956" by M. J. Neufeld 2006. *Space Policy*, vol. 22, pp. 52–62.

23. *Wernher von Braun: Dreamer of Space, Engineer of War* by M. J. Neufeld 2007. New York: Alfred A. Knopf.

24. *This New Ocean: The Story of the First Space Age* by W. E. Burrows 1998. New York: Random House, p. 147.

25. *Challenge to Apollo: The Soviet Union and the Space Race, 1945–1974* by A. A. Siddiqi 2000. Washington, DC: NASA.

26. Johnson was a power player in the Senate and a vigorous proponent of the newly formed space agency. Of course, he also made sure that the biggest new NASA center was located in his home state. Johnson Space Center near Houston is the place where astronauts are trained; its moniker—Mission Control—alludes to its central role in space missions.

27. *NASA's Origins and the Dawn of the Space Age* by D. S. F. Portree 1998. Monographs in Aerospace History #10, NASA History Division, Washington, DC.

28. The Space Act and its history of legislative amendments since 1958 can be found at the NASA History Office website, online at http://history.nasa.gov/spaceact-legishistory.pdf.

3: SEND IN THE ROBOTS

1. *The Race: The Uncensored Story of How America Beat Russia to the Moon* by J. Schefter 1999. New York: Doubleday. It was by no means smooth sailing initially. In the year after Sputnik, the Soviets successfully launched Sputnik 2 with the dog Laika aboard but failed twice to launch Sputnik 3, while the Americans launched their first satellite, Explorer 1, and also launched Vanguard 1, but they failed with five other Vanguard launches.

2. *The Rocket Men: Vostok and Voskhod, the First Soviet Manned Spaceflights* by R. Hall and D. J. Shayler 2001. New York: Springer-Praxis Books, pp. 149–55.

3. Gagarin attained worldwide celebrity after his feat, although he never flew in space again. With a warm personality and a megawatt smile, he attracted crowds wherever he went. But fame took its toll on him and he became an alcoholic. Conspiracy theories swirled around his death in a routine training flight in 1968, but it seems to have been caused when he was flying at low altitude and his fighter jet was caught in the wake of another fighter jet. "Yuri's Night" is celebrated in hundreds of cities around the world every year on April 12, the anniversary of his flight and of the first Space Shuttle mission.

4. "Special Message to the Congress on Urgent National Needs," a speech by Presi-

dent John F. Kennedy to a joint session of Congress on May 25, 1961, online at http://history.nasa.gov/moondec.html.

5. Many books have been written about the Apollo program. Two of the best are: *Apollo: The Race to the Moon* by C. Murray and C. B. Cox 1999. New York: Simon & Schuster; and *Moonshot: The Inside Story of Mankind's Greatest Adventure* by D. Parry 2009. Chatham, UK: Ebury Press. For two insider perspectives, see: *Failure Is Not an Option: Mission Control from Mercury to Apollo 13 and Beyond* by G. Kranz 2000. New York: Simon & Schuster; and *In the Shadow of the Moon: A Challenging Journey to Tranquility, 1965–1969* by F. French and C. Burgess 2007. Lincoln: University of Nebraska Press.

6. *John F. Kennedy and the Race to the Moon* by J. M. Logsdon 2010. New York: Palgrave Macmillan.

7. *In the Cosmos: Space Exploration and Soviet Culture* by J. T. Andrews and A. A. Siddiqi 2011. Pittsburgh: University of Pittsburgh Press.

8. *A Challenge to Apollo: The Soviet Union and the Space Race, 1945–1974* by A. A. Siddiqi 2000. Special Publication NASA-SP-2000-4408, Government Printing Office, Washington, DC.

9. *Apollo Expeditions to the Moon*, ed. by E. M. Cortright 1975. Special Publication NASA-SP-350, online at http://history.nasa.gov/SP-350/ch-11-4.html.

10. In particular, the public imagination was captured by "Earthrise," the image of the Earth rising over the lunar landscape, taken by William Anders during the Apollo 8 mission in 1968. Nature photographer Galen Rowell called it "the most influential environmental photograph ever taken."

11. "Animals as Cold Warriors: Missiles, Medicine, and Man's Best Friend," article at the US National Library of Medicine website, online at http://www.nlm.nih.gov/exhibition/animals/laika.html.

12. Quote from a news conference in Moscow, after his retirement in 1998, reported fourteen years later online at http://web.archive.org/web/20060108184335/http://www.dogsinthenews.com/issues/0211/articles/021103a.htm.

13. It's reasonable to question how much money is spent on something as seemingly esoteric as the space program, but NASA is almost literally a drop in the federal bucket. NASA's budget is close to $18 billion, or 15 cents a day for every American. That's forty times less than annual military spending. In terms of other things on which Americans spend money, it's thirty times less than gambling, three times less than spending on pets, and two times less than spending on pizza. If everyone would hold the pepperoni, we could send out more space missions.

14. *The Sidereal Messenger* by G. Galilei 1610 is a short pamphlet containing his observations of the Moon, the moons of Jupiter, and the Milky Way. The original is a very rare book, worth hundreds of thousands of dollars, but a commentary was published in 2010 in *Isis*, vol. 101, no. 3, pp. 644–45.

15. The learning curve is apparent in the success rate of missions to the inner Solar System: the Moon, Mars, and Venus. Using tabulations on Wikipedia (NASA infor-

mation is too chaotically organized to do it using their websites), the success rate of space probes went from 65 percent in the 1960s to 73 percent in the 1970s to 87 percent in the 1980s. It took a downtick to 72 percent in the 1990s and then improved again to 91 percent in the 2000s.

16. *Pale Blue Dot: A Vision of the Human Future in Space* by C. Sagan 1994. New York: Random House, pp. xv–xvi.

17. By the time of its final flight in 2011, the Space Shuttle had served fifteen years longer than the time for which it had been designed. After a call for proposals from museums and public institutions, NASA distributed the four remaining orbiters: original Shuttles Atlantis and Discovery, Challenger's replacement Endeavor, and an atmospheric test orbiter named Enterprise. Kennedy Space Center, the Smithsonian National Air and Space Museum, the California Science Center, and the Intrepid Sea-Air-Space Museum in New York City were the lucky recipients.

18. After the Challenger disaster, President Reagan formed the Rogers Commission to investigate. In their televised hearings, physicist Richard Feynman had a memorable moment when he dipped an O-ring into a cup of ice water to show how it became less resilient at the low temperatures at the time of launch. He was scathing about the wildly unrealistic estimates of reliability from NASA engineers and the stark failures of NASA management: "For a successful technology, reality must take precedence over public relations, for nature cannot be fooled." Rogers Commission Report 1986, Appendix F.

4: REVOLUTION IS COMING

1. NASA still attracts talented scientists and engineers. I've taught more than 800 engineers at six NASA centers, and most of them feel zeal and passion for their work. But the agency's ability to attract the best and the brightest peaked during the Apollo era. In the 1970s and 1980s, the lure of Silicon Valley proved stronger; in the 1990s and 2000s, the rise of the Internet and the "dot-com boom" offered a new frontier with no apparent limits. Like any government agency, NASA has layers of bureaucracy and the culture can often be far from entrepreneurial.

2. *NASA's Efforts to Reduce Unneeded Infrastructure and Facilities* 2013, Report Number IG-13-008, Office of the Inspector General, Washington, DC.

3. *Final Countdown: NASA and the End of the Space Shuttle Program* by P. Duggins 2007. Tampa: University of Florida Press.

4. As we've seen, Russia had more than its fair share of space pioneers and visionaries. The technical education provided by major Russian universities was unparalleled, even as the country limped along with decrepit infrastructure and uncompetitive industries. Russian scientists and engineers were treated well and provided with perks that made life tolerable. But all that changed with the chaos that followed the 1989 fall of the Soviet Union. Since then, universities have been starved of

resources and Russia has been suffering a severe brain drain that has sent much of its technical talent to the United States and Western Europe.

5. As reported by National Public Radio in a 2012 story on mounting problems in the Soviet space program, online at http://www.npr.org/2012/03/12/148247197/ for-russias-troubled-space-program-mishaps-mount.

6. NASA's budget hovers around $18 billion, and it hasn't changed much in real terms in more than a decade. At the moment, the pie is roughly divided into 28 percent for Earth and space science and astrophysics, 22 percent for the development of rockets and propulsion systems, and 22 percent for the International Space Station, with the rest for aeronautics and other technology development.

7. "The Interplanetary Internet" by J. Jackson 2005, published by the online magazine of the IEEE, at http://spectrum.ieee.org/telecom/internet/the-interplanetary -internet.

8. A mesmerizing video showing Baumgartner's perspective as he plunged 24 miles has had more than five million views on YouTube: https://www.youtube.com/ watch?v=raiFrxbHxV0. In 2014, Google VP Alan Eustace broke Baumgartner's altitude record (though not his speed record) by falling and parachuting from 135,890 feet to the ground in just fifteen minutes.

9. The story of the competition between astronauts and test pilots to reach space was well told by Tom Wolfe in his book *The Right Stuff* 1979. New York: Farrar, Straus and Giroux. The Mercury 7 astronauts were not originally intended to fly their space capsules, and Wolfe contrasted their roles with the high skill of test pilots such as Chuck Yeager, who took their planes to the edge of space. Wolfe's book was made into a popular film in 1983.

10. *Chuck Yeager and the Bell X-1: Breaking the Sound Barrier* by D. A. Pisano, F. R. van der Linden, and F. H. Winter 2006. Washington, DC: Smithsonian National Air and Space Museum.

11. *Press On! Further Adventures in the Good Life* by C. Yeager and C. Leerhsen 1997. New York: Bantam Books.

12. *At the Edge of Space: The X-15 Flight Program* by M. O. Thompson 1992. Washington and London: Smithsonian Institution Press.

13. The US Air Force considered an altitude of 50 miles or 80 kilometers to be the limit of space. However, the Fédération Aéronautique Internationale, the world governing body of aviation records, set the limit of space at 62 miles or 100 kilometers; by that criterion, only one Air Force pilot was truly an astronaut.

14. *American X-Vehicles: An Inventory from X-1 to X-50* by D. R. Jenkins, T. Landis, and J. Miller 2003, Monographs in Space History (Centennial of Flight), NASA Special Publication Number 31, NASA History Office, Washington, DC.

15. Space poetry is a small niche. Several collections of poetry about science include poems inspired by the space program, particularly Apollo and journeys to the Moon, most notably *Songs from Unsung Worlds: Science in Poetry* 1988, ed. by Bonnie Gordon. London: Birkhäuser; and *Contemporary Poetry and Contem-*

porary Science 2006, ed. by Robert Crawford. Oxford: Oxford University Press. Amazingly, it wasn't until 2009 that an astronaut penned a poem while in orbit. The honor went to American Don Pettit for "Halfway to Pluto." Also in that year, Japanese astronaut Koichi Wakata wrote a stanza as part of a free-association chain poem based on venerable Japanese *renga* and *renku* forms.

16. *Astronaut Fact Book* 2013, NASA Publication NP-2013-04-003-JSC, National Aeronautics and Space Administration, Washington, DC.

17. *The Colbert Report*, Episode 1012, broadcast on November 3, 2005, on Comedy Central. See episodes online at http://www.thecolbertreport.cc.com.

18. "Prospects of Space Tourism" by S. Abitzsch 1996, presented at the Ninth European Aerospace Congress, hosted by Space Future.

19. *Space Tourism: Do You Want to Go?* by J. Spencer 2004. Burlington, Ontario: Apogee Books.

20. The swift death of the program at the hands of the US House of Representatives is recounted on the NASA Watch website: http://nasawatch.com/archives/2005/06/nasas-first-and-last-artist-in-residence.html.

21. *Inventing the Internet* by J. Abbate 1999. Cambridge: MIT Press. Also see: "The Internet: On its International Origins and Collaborative Vision" by R. Hauben 2004. *Amateur Computerist*, vol. 2, no. 2, and "A Brief History of the Internet" by B. M. Leiner et al. 2009, online at http://www.internetsociety.org/internet/what-internet/history-internet/brief-history-internet.

22. "Eisenhower's Warning: The Military-Industrial Complex Forty Years Later" by W. D. Hartung 2001. *World Policy Journal*, vol. 18, no. 1.

23. *Unwarranted Influence: Dwight D. Eisenhower and the Military-Industrial Complex* by J. Ledbetter 2011. New Haven, CT: Yale University Press.

5: MEET THE ENTREPRENEURS

1. "Private Space Exploration a Long and Thriving Tradition" by M. Burgan 2012. In *Bloomberg View*, online at http://www.bloombergview.com/articles/2012-07-18/private-space-exploration-a-long-and-thriving-tradition.

2. "The Wit and Wisdom of Burt Rutan" by E. R. Hedman 2011. In *The Space Review*, online at http://www.thespacereview.com/article/1910/1.

3. *Rutan—The Canard Guru* by M. S. Rajamurthy 2009. India: National Aerospace Laboratories.

4. *Voyager* by J. Yeager and D. Rutan 1988. New York: Alfred A. Knopf. Also: *Voyager: The World Flight: The Official Log, Flight Analysis and Narrative Explanation* by J. Norris 1988. Northridge, CA: Jack Norris.

5. "Burt Rutan—Aerospace Engineer," interview on March 3, 2012, on BigThink website, http://bigthink.com/users/burtrutan.

6. The excitement of winning the X Prize was captured in the documentary *Mojave*

Magic: A Turtle's Eye View of SpaceShipOne. This 2005 short film was directed and written by Jim Sayers, produced by Dag Gano and Jim Sayers, and distributed by Desert Turtle Productions.

7. *Losing My Virginity: How I've Survived, Had Fun, and Made a Fortune Doing Business My Way* by R. Branson 2002. London: Virgin Books Limited. See also *Screw Business As Usual* by R. Branson 2011. London: Penguin Group.

8. "Richard Branson: Virgin Entrepreneur" by M. Vinnedge 2009. *Success* magazine, online at http://www.success.com/article/richard-branson-virgin-entrepreneur.

9. *Dirty Tricks: The Inside Story of British Airways' Secret War Against Richard Branson's Virgin Atlantic* by M. Gregory 1994. London: Little, Brown.

10. "Up: The Story Behind Richard Branson's Goal to Make Virgin a Galactic Success" by A. Higginbotham 2013. *Wired* magazine, online at http://www.wired.co.uk/magazine/archive/2013/03/features/up.

11. From a Reddit discussion on October 17, 2013, online at http://www.reddit.com/r/IAmA/comments/1onkop/i_am_peter_diamandis_founder_of_xprize/.

12. *We* by C. Lindbergh 1927. New York: Putnam and Sons. The title refers to the fact that Lindbergh never referred to himself in making his historic flight—he always twinned himself with his plane, the *Spirit of St. Louis.*

13. "The Dream of the Medical Tricorder" 2012. *The Economist,* online at http://www.economist.com/news/technology-quarterly/21567208-medical-technology-hand-held-diagnostic-devices-seen-star-trek-are-inspiring.

14. "Peter Diamandis: Rocket Man" by B. Caulfield 2012. *Forbes* magazine, February 13, online at http://www.forbes.com/sites/briancaulfield/2012/01/26/peter-diamandis-rocket-man/2/.

15. Diamandis recounts the story of Hawking's zero-gravity ride in his blog entry for February 15, 2013, in the *Huffington Post,* online at http://www.huffingtonpost.com/peter-diamandis/prof-hawking-goes-weightl_b_2696167.html.

16. "Robert Goddard: A Man and His Rocket," online at http://www.nasa.gov/missions/research/f_goddard.html.

17. *Abundance: The Future Is Better Than You Think* by P. Diamandis and S. Kotler 2012. New York: Free Press.

18. Quoted in "The New Space Race: Complicating the Rush to the Stars" by D. Bennett for the *Tufts Observer,* online at http://tuftsobserver.org/2013/11/the-new-space-race-complicating-the-rush-to-the-stars/.

19. "At Home with Elon Musk: The (Soon-to-Be) Bachelor Billionaire" by H. Elliott in *Forbes Life,* online at http://www.forbes.com/sites/hannahelliott/2012/03/26/at-home-with-elon-musk-the-soon-to-be-bachelor-billionaire/.

20. *The Startup Playbook: Secrets of the Fastest-Growing Startups from Their Founding Entrepreneurs* by D. Kidder 2013. San Francisco: Chronicle Books.

21. See *The Economist,* online at http://www.economist.com/news/technology-quarterly/21603238-bill-stone-cave-explorer-who-has-discovered-new-things-about-earth-now-he.

22. *Born Entrepreneurs, Born Leaders: How Your Genes Affect Your Work Life* by S. Shane 2010. Oxford: Oxford University Press. See also the technical article "Is the Tendency to Engage in Entrepreneurship Genetic?" by N. Nicolaou, S. Shane, L. Cherkas, J. Hunkin, and T. D. Spector 2008. *Management Science*, vol. 54, no. 1, pp. 167–79.

23. "The Innovative Brain" by A. Lawrence, L. Clark, J. N. Labuzetta, B. Sahakian, and S. Vyakarnum 2008. *Nature*, vol. 456, pp. 168–69.

6: BEYOND THE HORIZON

1. *The Heavens and the Earth: A Political History of the Space Age* by W. MacDougall 1985. Baltimore: Johns Hopkins University Press.

2. Quoted in "Private Dragon Capsule Arrives at Space Station in Historic First" by C. Moskowitz. Space.com, online at http://www.space.com/15874-private-dragon-capsule-space-station-arrival.html.

3. There was a sobering reminder of the difficulties of spaceflight when an Antares rocket built by Orbital Sciences exploded seconds after launch on October 28, 2014. The rocket was on a resupply mission to the International Space Station.

4. Quoted in "SpaceX Successfully Launches Its Next Generation Rocket" by A. Knapp. *Forbes* magazine, online at http://www.forbes.com/sites/alexknapp/2013/09/30/spacex-successfully-launches-its-next-generation-rocket/.

5. Paris Hilton, quoted in Britain's *Daily Express*, online at http://www.express.co.uk/news/science-technology/431046/Hollywood-stars-in-space-as-Richard-Branson-s-Earth-orbiting-flight-is-months-away.

6. Space Adventures was one of the first players in the commercial space sector. Eric Anderson founded the company in 1998; since 2001, he's sent seven clients on eight successful trips to the International Space Station. Among those who have placed $5 million deposits on future orbital spaceflights is Sergey Brin, the cofounder of Google. The company plans to start offering flybys of the Moon in 2017, at a cost of $150 million per trip.

7. Quoted in "Amazon.com's Bezos Invests in Space Travel, Time" by Amy Martinez, in *Seattle Times*, online at http://seattletimes.com/html/businesstechnology/2017883721_amazonbezos25.html.

8. The Commercial Orbital Transportation Services (COTS) program was NASA's effort to develop America's capacity to resupply the International Space Station. The program ran from January 2006 to September 2013 and awarded $500 million to SpaceX and Orbital Sciences Corporation. This is less than the cost of a single Space Shuttle flight, so NASA considers the program a great success. See *Commercial Orbital Transportation Services: A New Era in Spaceflight* by R. Hackler and R. Wright 2014, NASA Special Publication 2014-017, NASA, Washington, DC.

9. NASA hosts information online at http://www.nasa.gov/exploration/systems/sls/#
   .U5Ot13JdWSo.

10. For a transcript of the president's speech: http://www.nasa.gov/news/media/trans/
    obama_ksc_trans.html.

11. The asteroid initiative was originally dressed up with the idea of helping to protect
    the Earth from future impacts, but the rock that can be captured is far too small
    to be relevant to that problem. There are a lot of unknowns in the mission concept,
    and that typically translates into a great deal of unplanned time and money. See
    NASA's web page on the concept at http://www.nasa.gov/mission_pages/asteroids/
    initiative/.

12. "So You Want to Launch a Rocket? An Analysis of FAA Licensing Requirements
    with a Focus on the Legal and Regulatory Issues Created by the New Generation
    of Launch Vehicles," unpublished paper by Nathanael Horsley.

13. "Stuck to the Ground by Red Tape" 2013. *The Economist*, online at http://www
    .economist.com/news/technology-quarterly/21578517-space-technology-dozens
    -firms-want-commercialise-space-various-ways.

14. Burt Rutan testimony to the House Committee on Science, Subcommittee on
    Space and Aeronautics, hearing entitled "Future Markets for Commercial Space,"
    April 20, 2005.

15. The entire 2003 report is online at http://www.nasa.gov/columbia/home/CAIB_
    Vol1.html.

16. "Weighing the Risks of Space Travel" by J. Foust 2013. *The Space Review*, online
    at http://www.thespacereview.com/article/36/1.

17. As reported in an article on the 2001 book *Almost History* by R. Bruns, online at
    http://www.space.com/7011-president-nixon-prepared-apollo-disaster.html.

18. *The Evolution of Rocket Technology* (e-book) by M. D. Black 2012. Payloadz.com.

19. China has progressed rapidly in science and technology because of a buoyant econ-
    omy and strong investment. But a less noble reason is a shameless willingness to
    plagiarize intellectual property and to reverse engineer cutting-edge technologies.
    They have done this with fighter aircraft and supercomputers and they are now
    doing it with space technology.

20. "Space Transportation Costs: Trends in Price Per Pound to Orbit 1990–2000," a
    2002 report developed by the Futron Corporation in Bethesda, Maryland.

21. "The Effects of Long-Duration Space Flight on Eye, Head, and Trunk Coordi-
    nation During Locomotion" by I. B. Kozlovskaya et al. 2004. Unpublished report
    by Life Sciences Group, Johnson Space Center. Also, the National Academy of
    Sciences commissioned a report on adaptation to space, "Human Factors in Long-
    Duration Spaceflight," by the Space Sciences Board of the National Research
    Council 1972. Washington, DC: National Academies Press.

22. From the Johnson Space Center's oral history project, as interviewed in 1999 by
    Carol Butler, online at http://www.jsc.nasa.gov/history/oral_histories/StevensonRE/
    RES_5-13-99.pdf.

23. "Why Do Astronauts Suffer from Space Sickness?" An article on research by S. Nooij of Delft University of Technology, online at http://www.sciencedaily.com/releases/2008/05/080521112119.htm.

24. *Space Physiology and Medicine* by A. E. Nicogossian, C. L. Huntoon, and S. L. Pool 1993. Philadelphia: Lea and Febiger. See also "Beings Not Made for Space" by K. Chang, *New York Times*, January 27, 2014, online at http://www.nytimes.com/2014/01/28/science/bodies-not-made-for-space.html.

25. "Living and Working in Space," NASA Report FS-2006-11-030-JSC, produced by Johnson Space Center.

26. See the video segment from Comedy Central's *The Colbert Report* at http://www.colbertnation.com/the-colbert-report-collections/307748/colbert-s-best-space-moments/168719.

7: A PLETHORA OF PLANETS

1. Information about the Timbisha Shoshone Tribe can be found at the Death Valley National Park page of the National Park Service website, online at http://www.nps.gov/deva/parkmgmt/tribal_homeland.htm.

2. See the travelogue "Life in the Past Tense: Chile's Atacama Desert" by S. Beale, at the Perceptive Travel website: http://www.perceptivetravel.com/issues/1211/chile.html.

3. "Life Is a Chilling Challenge in Subzero Siberia" by B. Trivedi, from a National Geographic Channel TV show, online at http://news.nationalgeographic.com/news/2004/05/0512_040512_tvoymyakon.html.

4. *Pale Blue Dot: A Vision of the Human Future in Space* by C. Sagan 1994. New York: Random House, pp. xv–xvi.

5. "Extremophiles 2002" by M. Rossi et al. 2003. *Journal of Bacteriology*, vol. 185, no. 13, pp. 3683–89. See also *Polyextremophiles: Life Under Multiple Forms of Stress*, ed. by J. Seckbach et al. 2013. Dordrecht: Springer; and *Weird Life: The Search for Life That Is Very, Very Different from Our Own* by D. Toomey 2014. New York: W. W. Norton.

6. "Quick Guide: Tardigrades" by B. Goldstein and M. Baxter 2002. *Current Biology*, vol. 12, no. 14, R475; and "Radiation Tolerance in the Tardigrade" by D. D. Horikawa et al. 2006. *International Journal of Radiation Biology*, vol. 82, no. 12, pp. 843–48.

7. "The Role of Vitrification in Anhydrobiosis" by J. H. Crowe, J. F. Carpenter, and L. M. Crowe 1998. *Annual Review of Physiology*, vol. 60, pp. 73–103.

8. The simple calculation of the equilibrium temperature at the surface of a terrestrial planet depends only on the luminosity of the star and the distance to the planet. If the orbit is too eccentric or elliptical, a planet may move in and out of the habitable zone over the course of its year. Atmospheric gases raise the surface

temperature—strongly in the case of greenhouse gases like carbon dioxide and methane—and that shifts the habitable zone out to larger distances.

9. "A Possible Biogeochemical Model for Mars" by A. De Morais 2012. *43rd Lunar and Planetary Science Conference*, vol. 43, p. 2943.

10. Abel Mendez at the University of Puerto Rico has used quantitative measures of planetary habitability to evaluate locations in the Solar System. The measures are keyed to the survival of primary producers like plants, phytoplankton, and microbes in general. Habitability can and will evolve as the atmospheric and geological conditions change. In the Solar System, Mendez found Enceladus to have the best subsurface habitability, followed by Mars, Europa, and Titan. For a summary, see the *Astrobiology* magazine article online at http://www.astrobio.net/pressrelease/3270/islands-of-life-across-space-and-time.

11. "A Jupiter-Mass Companion to a Solar-Type Star" by M. Mayor and D. Queloz 1995. *Nature*, vol. 378, pp. 355–59.

12. *The Exoplanet Handbook* by M. A. C. Perryman 2011. Cambridge: Cambridge University Press.

13. "The HARPS Search for Earth-like Planets in the Habitable Zone" by F. Pepe et al. 2011. *Astronomy and Astrophysics*, vol. 534, p. A58.

14. "One or More Bound Planets per Milky Way Star from Microlensing Observations" by A. Cassan et al. 2012. *Nature*, vol. 481, pp. 167–69; and "Prevalence of Earth-Size Planets Orbiting Sun-like Stars" by E. A. Petigura, A. W. Howard, and G. W. Marcy 2013. *Proceedings of the National Academy of Sciences*, vol. 110, no. 48, p. 19273.

15. A large amount of technical and scientific information about the Kepler mission and its goals is on the NASA website: http://www.kepler.arc.nasa.gov/.

16. In 2014, Kepler got a new lease on life when engineers figured out how to use the pressure of starlight and tiny thruster burns to maintain the spacecraft orientation. It's not as precise as the original pointing, so Kepler can no longer detect Earths, but it can find planets around a wide range of star types over a wide swath of sky. The so-called K2 mission will last about two years.

17. "Planetary Candidates Observed by Kepler, III. Analysis of the First 16 Months of Data" by N. Batalha et al. 2013. *The Astrophysical Journal Supplement*, vol. 204, pp. 24–45.

18. "The Occurrence Rate of Small Planets Around Small Stars" by C. D. Dressing and D. Charbonneau 2013. *The Astrophysical Journal*, vol. 767, pp. 95–105.

19. "Space Oddities: 8 of the Strangest Exoplanets" by D. Orf 2013. *Popular Mechanics* magazine, online at http://www.popularmechanics.com/science/space/deep/space-oddities-8-of-the-strangest-exoplanets#slide-1.

20. "An Earth Mass Planet Orbiting Alpha Centauri B" by X. Dumusque et al. 2012. *Nature*, vol. 491, pp. 207–11.

8: THE NEXT SPACE RACE

1. "Profiles of Government Space Programs 2014" published by Euroconsult, with a summary analysis online at http://spaceref.biz/commercial-space/global-spending-on-space-decreases-for-first-time-in-20-years.html.

2. "China: The Next Space Superpower" by E. Strickland 2013, a detailed analysis for *IEEE Spectrum* magazine, online at http://spectrum.ieee.org/aerospace/space-flight/china-the-next-space-superpower.

3. *China's Space Program: From Conception to Manned Spaceflight* by B. Harvey 2004. Dordrecht: Springer-Verlag.

4. Hadfield's video went viral on YouTube, attracting more than 22 million views. But he only had agreement from David Bowie to keep it up for a year, so it was taken down in May 2014.

5. Reported by Space Daily, online at http://www.spacedaily.com/reports/China_launches_longest-ever_manned_space_mission_999.html.

6. "Chinese Super-Heavy Launcher Designs Exceed Saturn V" by B. Perrett 2013. *Aviation Week*, online at http://www.aviationweek.com/Article.aspx?id=/article-xml/AW_09_30_2013_p22-620995.xml.

7. Online reporting from Space.com, at http://www.space.com/14697-china-space-program-military-threat.html, and http://www.space.com/25517-china-military-space-technology.html.

8. "The Man Who Says He Owns the Moon" by R. Hardwick, article and interview on *Motherboard*, online at http://motherboard.vice.com/blog/the-man-who-owns-the-moon.

9. The full text of the treaty, in English, French, Russian, Spanish, Chinese, and Arabic, is on the United Nations website at http://www.unoosa.org/oosa/SpaceLaw/outerspt.html.

10. The committee was formed in 1959 by the UN General Assembly. There are seventy-six member states taking part, and it held its 57th session in 2014 in Vienna. There are two standing subcommittees: Scientific and Technical, and Legal. The website is http://www.unoosa.org/oosa/COPUOS/copuos.html.

11. The UN Moon Treaty entered into force after being ratified by five countries in 1984. The full text is online at http://www.unoosa.org/oosa/SpaceLaw/moon.html.

12. "Is NASA's Plan to Lasso an Asteroid Really Legal?" by L. David 2013, at the Space.com website: http://www.space.com/22605-nasa-asteroid-capture-mission-legal-issues.html.

13. Quoted in the article "To the Moon, Mars, and Beyond: Culture, Law, and Ethics in Space-Faring Societies" by L. Billings, presented in 2006 at the 21st annual conference of the International Association for Science, Technology, and Society.

14. Michael Griffin, quoted by Linda Billings in chapter 25 of *Societal Impact of Spaceflight*, ed. by S. J. Dick and R. A. Launius, NASA Special Publication

NASA-SP-4801, National Aeronautics and Space Administration, Washington, DC. Peter Diamandis is quoted in "The Final Capitalist Frontier" by M. Baard, in *Wired* magazine, online at http://www.wired.com/science/space/news/2004/11/65729.

15. "The Space Elevator: A Thought Experiment or the Key to the Universe?" by A. C. Clarke, in *Advances in Earth Oriented Applied Space Technologies*, Vol. 1, 1981. London: Pergamon Press, pp. 39–48. See also "The Physics of the Space Elevator" by P. K. Aravind 2007. *American Journal of Physics*, vol. 45, no. 2, p. 125.

16. The state of the art just after nanotubes were developed was given in "Space Elevators: An Advanced Earth-Space Infrastructure for the New Millennium," compiled by D. B. Smitherman Jr., NASA Publication CP-2000-210429, based on findings from the Advanced Space Infrastructure Workshop on Geostationary Orbiting "Space Elevator" Tether Concepts, held at NASA's Marshall Space Flight Center in June 1999. Since that time, Bradley Carl Edwards has pursued the carbon nanotube route to a space elevator with the support of NASA's Institute for Advanced Concepts.

17. As with suborbital flight, progress has been spurred by a series of competitions similar to the Ansari X Prize. *Elevator: 2010* ran challenges every year from 2005 to 2009 and NASA raised the prize money, using its Centennial Challenges program. Meanwhile, the Europeans started their own competition in 2011.

18. "Carbyne from First Principles: Chain of C Atoms, a Nanorod, or a Nanorope?" by M. Liu et al. 2013. *American Chemical Society Nanotechnology*, vol. 7, no. 11, pp. 10075–82.

19. *Space Elevators: An Assessment of the Technological Feasibility and the Way Forward* by P. Swan et al. 2013. Houston: Science Deck Books, Virginia Edition Publishing Company.

20. "The Economic Benefits of Commercial GPS Use in the United States and the Costs of Potential Disruption," by N. D. Pham, June 2011, NDP Consulting, online at http://www.saveourgps.org/pdf/GPS-Report-June-22-2011.pdf.

21. "The Economic Impact of Commercial Space Transportation on the U.S. Economy in 2009," a 2010 report by the Federal Aviation Administration's Office of Commercial Space Transportation.

22. "Space Tourism Market Study: Orbital Space Travel and Destinations with Suborbital Space Travel," an October 2002 report by the Futron Corporation, Bethesda, Maryland. A more recent report by the FAA, "Suborbital Reusable Vehicles: A Ten-Year Forecast of Market Demand," reaches similar conclusions.

23. *Mining the Sky: Untold Riches from the Asteroids, Comets, and Planets* by J. S. Lewis 1998. New York: Basic Books.

24. "Orbit and Bulk Density of the OSIRIS-REx Target Asteroid (101955) Bennu" by S. R. Chesley et al. 2014. *Icarus*, vol. 235, pp. 5–22.

25. "Profitable Asteroid Mining" by M. Busch 2004. *Journal of the British Interplanetary Society*, vol. 57, pp. 301–5.

9: OUR NEXT HOME

1. The working group's deliberations are described in the epilogue to "Chariots for Apollo: A History of Manned Lunar Spacecraft" by C. G. Brooks, J. M. Grimwood, and L. S. Swenson 1979, published as NASA Special Publication 4205 in the NASA History Series.

2. "Costs of an International Lunar Base" by J. Weppler, V. Sabathier, and A. Bander 2009, Center for Strategic and International Studies, Washington, DC, online at https://csis.org/publication/costs-international-lunar-base.

3. "How Wet the Moon? Just Damp Enough to Be Interesting" by R. A. Kerr 2010. *Science*, vol. 330, p. 434, and a subsequent set of research articles in the special issue of *Science*.

4. "Mining and Manufacturing on the Moon," from the Aerospace Scholars program, online at http://web.archive.org/web/20061206083416/http://aerospacescholars .jsc.nasa.gov/HAS/cirr/em/6/6.cfm; and "Building a Lunar Base with 3D Printing," a research program at the European Space Agency, online at http://www.esa.int/ Our_Activities/Technology/Building_a_lunar_base_with_3D_printing.

5. "The Peaks of Eternal Light on the Lunar South Pole: How They Were Found and What They Look Like" by M. Kruijff 2000. *4th International Conference on Exploration and Utilisation of the Moon* (ICEUM4), ESA/ESTEC, SP-462. Also: "A Search for Lava Tubes on the Moon: Possible Lunar Base Habitats" by C. R. Coombs and B. R. Hawke 1992. *Second Conference on Lunar Bases and Space Activities of the 21st Century* (SEE N93-17414 05-91), vol. 1, pp. 219–29.

6. "Lunar Space Elevators for Cislunar Space Development" by J. Pearson, E. Levin, J. Oldson, and H. Wykes 2005, Phase 1 Final Technical Report under research subaward 07605-003-034, submitted to NASA.

7. Information is routinely updated online at http://www.googlelunarxprize.org/.

8. "Estimation of Helium-3 Probable Reserves in Lunar Regolith" by E. N. Slyuta, A. M. Abdrakhimov, and E. M. Galimov 2007. *Lunar and Planetary Science Conference XXXVIII*, pp. 2175–78.

9. "Nuclear Fusion Energy—Mankind's Giant Step Forward" by S. Lee and L. H. Saw 2010. *Proceedings of the Second International Conference on Nuclear and Renewable Energy Sources*, pp. 2–8.

10. "China Considers Manned Moon Landing Following Breakthrough Chang'e-3 Mission Success" by K. Kremer, as reported in *Universe Today*, online at http:// www.universetoday.com/107716/china-considers-manned-moon-landing-following -breakthrough-change-3-mission-success/.

11. *The War of the Worlds* by H. G. Wells 1898. London: Bell, quote from chapter 1.

12. "Metastability of Liquid Water on Mars" by M. H. Hecht 2002. *Icarus*, vol. 156, pp. 373–86. Also: "Ancient Oceans, Ice Sheets, and the Hydrological Cycle on Mars" by V. R. Baker et al. 1991. *Nature*, vol. 352, pp. 589–94. More recent discoveries

are covered in "Introduction to Special Issue: Analysis of Surface Materials by the Curiosity Mars Rover" by J. Grotzinger 2013. *Science*, vol. 341, p. 1475, and subsequent articles in the special issue of *Science* magazine.

13. *Water on Mars* by M. H. Carr 1996. Oxford: Oxford University Press.

14. *The Case for Mars: The Plan to Settle the Red Planet and Why We Must* by R. M. Zubrin and R. Wagner 1996. New York: Simon & Schuster; *Mars on Earth: The Adventures of Space Pioneers in the High Arctic* by R. M. Zubrin 2003. New York: Bargain Books; *How to Live on Mars: A Trusty Guidebook to Surviving and Thriving on the Red Planet* by R. M. Zubrin 2008. New York: Three Rivers Press. His most recent book brings Mars exploration up to date with the Mars Direct proposal using the DragonX rocket: *Mars Direct, Space Exploration, and the Red Planet* by R. M. Zubrin 2013. New York: Penguin.

15. NPR *Science Friday* interview, online at http://www.npr.org/2011/07/01/137555244/is-settling-mars-inevitable-or-an-impossibility.

16. *Pathways to Exploration: Rationales and Approaches for a U.S. Program of Human Space Exploration*, by the Committee on Human Spaceflight 2014, National Research Council, Washington, DC.

17. "Circadian Rhythm of Autonomic Cardiovascular Control During Mars 500 Simulated Mission to Mars" by D. E. Vigo et al. 2013. *Aviation and Space Environmental Medicine*, vol. 84, pp. 1023–38.

18. From a post by Buzz Aldrin on his website: http://buzzaldrin.com/what-nasa-has-wrong-about-sending-humans-to-mars/.

19. The website of Inspiration Mars, a 501(c)(3) nonprofit organization, is at http://www.inspirationmars.org/.

20. For more information, see: Inspiration Mars online at http://www.inspirationmars.org/ and Mars One online at http://www.mars-one.com/.

21. While the launch date has receded to 2024 at the earliest, the media angle is moving full steam ahead. In June 2014, Lansdorp inked a deal with the Dutch TV giant Endemol to start filming a reality series based on the training and culling of the final set of space travelers.

22. *Reading the Rocks: The Autobiography of the Earth* by M. Bjornerud 2005. New York: Basic Books; and *Life on a Young Planet: The First Three Billion Years of Evolution on Earth* by A. H. Knoll 2004. Princeton, NJ: Princeton University Press.

23. "Technological Requirements for Terraforming Mars" by R. M. Zubrin and C. P. McKay 1993, technical report for NASA Ames Research Center, online at http://www.users.globalnet.co.uk/~mfogg/zubrin.htm.

24. *Red Mars* by K. S. Robinson 1993 covers colonization; the quote in the following paragraph is from p. 171. *Green Mars* by K. S. Robinson 1994 covers terraforming. *Blue Mars* by K. S. Robinson 1995 covers the long-term future of human habitation. All are published by Random House (New York).

10: REMOTE SENSING

1. "Why Oculus Rift Is the Future of Gaming," online at http://www.gizmoworld.org/why-oculus-rift-is-the-future-in-gaming/.

2. Intriguingly, telepresence doesn't have to convey the remote scene with perfect fidelity, because the brain has a tendency to "fill in the blanks" and "smooth out the rough edges" of any representation that is familiar. See "Another Look at 'Being There' Experiences in Digital Media: Exploring Connections of Telepresence with Mental Imagery" by I. Rodriguez-Ardura and F. J. Martinez-Lopez 2014. *Computers in Human Behavior*, vol. 30, pp. 508–18.

3. *Brother Assassin* by F. Saberhagen 1997. New York: Tor Books.

4. See http://www.ted.com/talks/edward_snowden_here_s_how_we_take_back_the_internet.

5. "Multi-Objective Compliance Control of Redundant Manipulators: Hierarchy, Control, and Stability" by A. Dietrich, C. Ott, and A. Albu-Schaffer 2013. *Proceedings of the 2013 IEEE/RSJ International Conference on Intelligent Robots and Systems*, Tokyo, pp. 3043–50.

6. *Human Haptic Perception*, ed. by M. Grunwald 2008. Berlin: Birkhäuser Verlag.

7. "Telepresence" by M. Minsky 1980, *Omni* magazine. The magazine is defunct, but the paper can be found online at http://web.media.mit.edu/~minsky/papers/Telepresence.html.

8. Feynman delivered his lecture at the American Physical Society meeting at Caltech on December 29, 1959. A transcript of the talk is online at http://www.zyvex.com/nanotech/feynman.html. He concluded his talk by posing two challenges and offering a prize of $1,000 for each one. His challenge to fabricate a tiny motor was won a year later by William McLellan. His second challenge was to fit the entire text of the *Encyclopædia Britannica* on the head of a pin. In 1985, a Stanford graduate student won the second challenge by reducing the first paragraph of Dickens's *A Tale of Two Cities* by a factor of 25,000.

9. Feynman reprised his idea after nanotechnology began to take off. "There's Plenty of Room at the Bottom" by R. P. Feynman 1992. *Journal of Microelectromechanical Systems*, vol. 1, pp. 60–66; and "Infinitesimal Machinery" by R. P. Feynman 1993. *Journal of Microelectromechanical Systems*, vol. 2, pp. 4–14.

10. "Prokaryotic Motility Structures" by S. L. Bardy, S. Y. Ng, and K. F. Jarrell 2003. *Microbiology*, vol. 149, part 2, pp. 295–304.

11. *Synergetic Agents: From Multi-Robot Systems to Molecular Robotics* by H. Haken and P. Levi 2012. Weinheim, Germany: Wiley-VCH. The book that started off the entire field was *Engines of Creation: The Coming Era of Nanotechnology* by E. Drexler 1986. New York: Doubleday.

12. "The Next Generation of Mars Rovers Could Be Smaller Than Grains of Sand" by B. Ferreira 2012, in *Popular Science*, online at http://www.popsci.com/technology/article/2012-07/why-next-gen-rovers-could-be-smaller-grain-sand.

13. Research at Goddard Space Flight Center: http://www.nasa.gov/centers/goddard/news/ants.html.

14. *Nanorobotics: Current Approaches and Techniques*, ed. by C. Mavroidis and A. Ferreira 2013. New York: Springer.

15. *From the Earth to the Moon* by J. Verne 1865. Paris: Pierre-Jules Hetzel.

16. *Solar Sails: A Novel Approach to Interplanetary Flight* by G. Vulpetti, L. Johnson, and L. Matloff 2008. New York: Springer.

17. The Cosmos 1 concept is described in "LightSail: A New Way and a New Chance to Fly on Light" by L. Friedman 2009. *The Planetary Report* (The Planetary Society, Pasadena), vol. 29, no. 6, pp. 4–9. After its initial failure, the project is transitioning to the use of CubeSats, described in *Small Satellites: Past, Present, and Future*, ed. by H. Helvajian and S. W. Janson 2008. El Segundo, CA: Aerospace Press.

18. Sunjammer was canceled after $21 million had been spent, due to problems encountered by the contractor L'Garde Inc. Congressional representative Dana Rohrabacher made the ironic comment: "We never seem to be able to afford these small technology development projects that can have potentially huge impacts . . . but we can find billions and billions of dollars to build a massive launch vehicle with no payloads, and no missions." He was referring to NASA's SLS heavy lift rocket.

19. "Nanosats Are Go!" in *The Economist* magazine, online at http://www.economist.com/news/technology-quarterly/21603240-small-satellites-taking-advantage-smart phones-and-other-consumer-technologies.

20. "NAIC Study of the Magnetic Sail" by R. Zubrin and A. Martin 1999 (slide presentation), online at http://www.niac.usra.edu/files/library/meetings/fellows/nov99/320Zubrin.pdf.

21. "Searching for Interstellar Communications" by G. Cocconi and P. Morrison 1959. *Nature*, vol. 184, pp. 844–46.

22. "The Drake Equation Revisited. Part 1," a retrospective by Frank Drake in *Astrobiology Magazine*, online at http://www.astrobio.net/index.php?option=com_retro spection&task=detail&id=610.

23. *SETI 2020: A Roadmap for the Search for Extraterrestrial Intelligence*, ed. by R. D. Ekers, D. Culler, J. Billingham, and L. Scheffer 2003. Mountain View, CA: SETI Press.

24. "Neuroanatomy of the Killer Whale (*Orcinus orca*) from Magnetic Resonance Images" by L. Marino et al. 2004. *The Anatomical Record Part A*, vol. 281A, no. 2, pp. 1256–63.

11: LIVING OFF-EARTH

1. "Biospherics and Biosphere 2, Mission One" by J. Allen and M. Nelson 1999. *Ecological Engineering*, vol. 13, pp. 15–29.

2. "Life Under the Bubble" by J. F. Smith 2010, from *Discover* magazine, online at http://discovermagazine.com/2010/oct/20-life-under-the-bubble#.UkvfALNsdOA.

3. Several Biospherians have written about their experience. See *Life Under Glass: The Inside Story of Biosphere 2* by A. Alling and M. Nelson 1993. Santa Fe: Synergetic Press; and *The Human Experiment: Two Years and Twenty Minutes Inside Biosphere 2* by J. Poynter 2006. New York: Thunder's Mouth Press. After the second failed experiment, the facility was taken over by Columbia University, which hoped to use it as a research station and "west campus." But urban students didn't flock to take classes there, so it passed from Columbia University to the University of Arizona in 2011. As a lab for studying the complex interplay of climate, soil chemistry, and flora and fauna, Biosphere is unrivaled, even if it can never be operated as a sealed and self-contained ecosystem. The project's deep-pocketed investor, Ed Bass, once intended to sell small-scale biospheres as a commercial proposition.

4. "Two Former Biosphere Workers Are Accused of Sabotaging the Dome," April 5, 1994, from the archives of the *New York Times*, online at http://www.nytimes.com/1994/04/05/us/two-former-biosphere-workers-are-accused-of-sabotaging-dome.html.

5. *Dreaming the Biosphere* by R. Reider 2010. Albuquerque: University of New Mexico Press.

6. "Calorie Restriction in Biosphere 2: Alterations in Physiologic, Hematologic, Hormonal, and Biochemical Parameters in Humans Restricted for a Two-Year Period" by R. Walford, D. Mock, R. Verdery, and T. MacCallum 2002. *The Journals of Gerontology, Series A*, vol. 57, no. 6, pp. B211–24.

7. "Coral Reefs and Ocean Acidification" by J. A. Kleypas and K. K. Yates 2009. *Oceanography*, December, pp. 108–17.

8. "Lessons Learned from Biosphere 2: When Viewed as a Ground Simulation/Analog for Long Duration Human Space Exploration and Settlement" by T. MacCallum, J. Poynter, and D. Bearden 2004. SAE Technical Paper, online at http://www.janepoynter.com/documents/LessonsfromBio2.pdf.

9. *Spacesuits: The Smithsonian National Air and Space Museum Collection* by A. Young 2009. Brooklyn: Power House Books.

10. "The Retro Rocket Look" from *The Economist*, online at http://www.economist.com/news/technology-quarterly/21603234-spacesuits-new-generation-outfits-astronauts-being-developed-although.

11. NASA's Ames Research Center in California commissioned the space-colony studies, and they hold the drawings and design studies. To show that the vision hasn't been abandoned, the website quotes Michael Griffin, former administrator of

NASA, as saying, "I know that humans will colonize the Solar System and one day go beyond." The project continues with an annual design study open to middle and high school students from anywhere in the world. The 2014 competition had 562 entries from 1,567 students in eighteen countries. Online at http://settlement.arc .nasa.gov/.

12. The grim fate of the Biosphere is a reminder that habitable zones evolve on long timescales. A planet is kept habitable by a symbiotic relationship between living organisms and the rocks and oceans, an insight first noted by James Lovelock, who originated the Gaia hypothesis. The fundamental drive of habitability is energy from the parent star. The Sun was 25 percent dimmer 3 to 3.5 billion years ago, when life on Earth was microbial and oxygenic photosynthesis had not yet evolved. In the future, as the Sun uses its nuclear fuel and settles into a more compact con-figuration, it will "burn hotter," so the Earth will not remain habitable for the full span that the Sun has nuclear fuel, another 4 to 4.5 billion years. Microbes that live inside rock or far below the surface of the ocean are immune to moderate changes in solar radiation since they live off geological energy. As the Earth becomes intol-erably hot, we'll have to develop Biospheres for the whole population—assuming humans persist that long.

13. Since 2012, the Doomsday Clock has stood at five minutes to midnight, uncomfort-ably close to disaster. The explanation associated with that judgment is worth quot-ing in full: "The challenges to rid the world of nuclear weapons, harness nuclear power, and meet the nearly inexorable climate disruptions from global warming are complex and interconnected. In the face of such complex problems, it is diffi-cult to see where the capacity lies to address these challenges. The political pro-cesses in place seem wholly inadequate to meet the challenges to human existence that we confront. . . . The potential for nuclear weapons use in regional conflicts in the Middle East, Northeast Asia, and particularly South Asia is also alarming. . . . Safer nuclear reactor designs need to be developed and built, and more stringent oversight, training, and attention are needed to prevent future disasters. . . . [T]he pace of change may not be adequate . . . to meet the hardships that large-scale disruption of the climate portends." See: http://thebulletin.org/press-release/it-5 -minutes-midnight. The Doomsday Clock and its timeline can be seen online at http://thebulletin.org/timeline.

14. Tsiolkovsky quote from a letter written in 1911, online in Russian at http://www.rf .com.ua/article/388. Sagan quote from *Pale Blue Dot*, p. 371. Niven is quoted by Arthur C. Clarke in "Meeting of the Minds: Buzz Aldrin Visits Arthur C. Clarke," reported by A. Chaikin on February 27, 2001, on Space.com. Hawking quote from a transcript of a video interview for *BigThink*, online at http://bigthink.com/videos/ abandon-earth-or-face-extinction.

15. Lansdorp is quoted in the article "Is Mars for Sale?" by A. Wills, for Mashable.com, online at http://mashable.com/2013/04/09/mars-land-ownership-colonization/.

16. "Space Settlement Basics" by A. Globus, NASA Ames Research Center website, at http://settlement.arc.nasa.gov/Basics/wwwwh.html.

17. From the collection *Tales of Ten Worlds* by A. C. Clarke 1962. New York: Harcourt Brace.

18. "What Do Real Population Dynamics Tell Us About Minimum Viable Population Sizes?" by C. D. Thomas 1990. *Population Biology*, vol. 4, no. 3, pp. 324–27.

19. *Bottleneck: Humanity's Impending Impasse* by W. R. Catton 2009. Xlibris.

20. "Biodiversity and Intraspecific Genetic Variation" by C. Ramel 1998. *Pure and Applied Chemistry*, vol. 70, no. 11, pp. 2079–84.

21. "The Grasshopper's Tale" by R. Dawkins, in *The Ancestor's Tale: A Pilgrimage to the Dawn of Life* 2004. Boston: Houghton Mifflin.

22. "Genetics and Recent Human Evolution" by A. R. Templeton 2007. *International Journal of Organic Evolution*, vol. 61, no. 7, pp. 1507–19. Other researchers think the bottleneck may have been more drawn out and not due to sudden environmental change, with numbers dropping as low as 2,000 for tens of thousands of years until the Late Stone Age. (See also note 23.)

23. "Population Bottlenecks and Pleistocene Human Evolution" by J. Hawks, K. Hunley, S. H. Lee, and M. Wolpoff 2000. *Molecular Biology and Evolution*, vol. 17, no. 1, pp. 2–22. The complete story of our evolution from Africa is in "Explaining Worldwide Patterns of Human Variation Using a Coalescent-Based Serial Founder Model of Migration Outward from Africa" by M. DeGiorgio, M. Jakobsson, and N. A. Rosenberg 2009. *Proceedings of the National Academies of Science*, vol. 106, no. 38, pp. 16057–62.

24. "Legacy of Mutiny on the Bounty: Founder Effect and Admixture on Norfolk Island" by S. Macgregor et al. 2010. *European Journal of Human Genetics*, vol. 18, no. 1, pp. 67–72. The Tristan da Cunha case study is reported in *Population Genetics and Microevolutionary Theory* by A. R. Templeton 2006. New York: John Wiley, p. 93. "Amish Microcephaly: Long-Term Survival and Biochemical Characterization" by V. M. Siu et al. 2010. *American Journal of Medical Genetics A*, vol. 7, pp. 1747–51.

25. "'Magic Number' for Space Pioneers Calculated," a report on the work of John Moore by D. Carrington, reported in *New Scientist*, online at http://archive.is/Xa8I.

26. *Interstellar Migration and the Human Experience*, ed. by B. R. Finney and E. M. Jones 1985. Berkeley: University of California Press.

27. *Do Androids Dream of Electric Sheep?* by P. K. Dick 1968. New York: Doubleday. The novel served as the primary source for the 1992 film *Blade Runner*, directed by Ridley Scott and starring Harrison Ford. Released to mixed reviews, it has since become a cult classic. The Roy Batty quote is in the penultimate scene in the director's cut of the film.

28. "Cyborgs and Space" by M. E. Clynes and N. S. Kline 1960. *Astronautics*, September, p. 27.

29. Harbisson founded the Cyborg Foundation in 2010 to help humans become cyborgs. His eyeborg has been enhanced such that he can perceive color saturation as well as 360 different hues. He has a flair for publicity, and a short documentary about him won a Grand Jury Prize at the 2012 Sundance Film Festival. He has done performance art converting colors into music, and his art focuses on the relationship between color and sound; he's also had experimental theater and dance performances. Using his eyeborg, he has created live sonic "portraits" of celebrities, including Leonardo DiCaprio, Al Gore, Tim Berners-Lee, James Cameron, Woody Allen, and Prince Charles. A 2013 *Huffington Post* article, "Hacking Our Senses," features his 2012 TED Global Talk, "I Listen to Color," and quotes him saying, "I don't feel that I'm using technology, I don't feel that I'm wearing technology, I feel that I am technology." The article is online at http://www.huffingtonpost .com/neil-harbisson/hearing-color-cyborg-tedtalk_b_3654445.html.

30. Warwick's most influential paper in 2003 begins: "From a cybernetics viewpoint, the boundaries between humans and machines become almost inconsequential." Published as "Cyborg Morals, Cyborg Values, Cyborg Ethics" by K. Warwick 2003. *Ethics and Information Technology*, vol. 7, pp. 131–37. See also: "Future Issues with Robots and Cyborgs" by K. Warwick 2010. *Studies in Ethics, Law, and Technology*, vol. 6, no. 3, pp. 1–20.

31. Article from 2012 in *The Verge*, online at http://www.theverge.com/2012/8/8/ 3177438/cyborg-america-biohackers-grinders-body-hackers.

32. Kevin Warwick quote is from an FAQ on his website, at http://www.kevinwarwick .com/. Francis Fukuyama quote is from "The World's Most Dangerous Ideas: Transhumanism" by F. Fukuyama 2004. *Foreign Policy*, vol. 144, pp. 42–43. Ronald Bailey 2004 rebuttal to Fukuyama is from Reason online at http://reason.com/ archives/2004/08/25/transhumanism-the-most-dangero. If you want to drink deep from the transhumanist Kool-Aid, see "Why I Want to Be Transhuman When I Grow Up" by N. Bostrom 2008, in *Medical Enhancement and Posthumanity*, ed. by B. Gordijn and R. Chadwick. New York: Springer, pp. 107–37.

12: JOURNEY TO THE STARS

1. This particular quote has engendered a lot of speculation and misattribution. For example, it can*not* be reliably attributed to baseball manager and purveyor of malapropisms Yogi Berra. It seems to originate in nineteenth-century Denmark and was used but not coined by physicist Niels Bohr. A detailed discussion is online at http://quoteinvestigator.com/2013/10/20/no-predict/.

2. See http://www.scientificamerican.com/article/pogue-all-time-worst-tech-predictions/; and http://www.informationweek.com/it-leadership/12-worst-tech-predictions-of-all -time/d/d-id/1096169.

3. See http://www.smithsonianmag.com/history/the-world-will-be-wonderful-in-the
-year-2000-110060404/?no-ist.

4. "An Earth Mass Planet Orbiting Alpha Centauri B" by X. Dumusque et al. 2012.
Nature, vol. 491, pp. 207–11. See also "The Exoplanet Next Door" by E. Hand
2012. *Nature*, vol. 490, p. 323.

5. "Possibilities of Life Around Alpha Centauri B" by A. Gonzales, R. Cardenas-
Ortiz, and J. Hearnshaw 2013. *Revista Cubana de Física*, vol. 30, no. 2, pp. 81–83.
The exoplanet simulation paper on the Alpha Centauri system is "Formation and
Detectability of Terrestrial Planets Around Alpha Centauri B" by J. M. Guedes et
al. 2008. *The Astrophysical Journal*, vol. 679, pp. 1582–87.

6. "Atmospheric Biomarkers on Terrestrial Exoplanets" by F. Selsis 2004. *Bulletin
of the European Astrobiology Society*, no. 12, pp. 27–40. See also: "Can Ground-
Based Telescopes Detect the Oxygen 1.27 Micron Absorption Feature as a Biomarker in Exoplanets?" by H. Kawahara et al. 2012. *The Astrophysical Journal*,
vol. 758, pp. 13–28; and "Deciphering Spectral Fingerprints of Habitable Exoplanets" by L. Kaltenegger et al. 2010. *Astrobiology*, vol. 10, no. 1, pp. 89–102.

7. "Exoplanetary Atmospheres" by N. Madhusudhan, H. Knutson, J. Fortney, and
T. Barman 2014, in *Protostars and Planets VI*, ed. by H. Buether, R. Klessen,
C. Dullemond, and Th. Henning. Tucson: University of Arizona Press.

8. "Detection of an Extrasolar Planet Atmosphere" by D. Charbonneau, T. M. Brown,
R. W. Noyes, and R. L. Gilliland 2001. *The Astrophysical Journal*, vol. 568, pp.
377–84.

9. Web pages on space propulsion and interstellar travel are maintained by NASA's
Glenn Research Center: http://www.nasa.gov/centers/glenn/technology/warp/scales
.html.

10. For an overview, see *Project Orion: The True Story of the Atomic Spaceship* by
G. Dyson 2002. New York: Henry Holt. The original paper was "On a Method of
Propulsion of Projectiles by Means of External Nuclear Explosions, Part 1," by C. J.
Everett and S. M. Ulam 1955. University of California Los Alamos Lab, unclas-
sified document archived at http://www.webcitation.org/5uzTHJfF7. For more
recent technical design work, see "Physics of Rocket Systems with Separated Rockets and Propellant" by A. Zuppero 2010, online at http://neofuel.com/optimum/.

11. "Reaching for the Stars: Scientists Examine Using Antimatter and Fusion to Propel
Future Spacecraft," April 1999, NASA, online at http://science1.nasa.gov/science
-news/science-at-nasa/1999/prop12apr99_1/.

12. The Rand Corporation, online at http://www.rand.org/pubs/research_memoranda/
RM2300.html.

13. "Galactic Matter and Interstellar Flight" by R. W. Bussard 1960. *Astronautica
Acta*, vol. 6, pp. 179–94.

14. "Roundtrip Interstellar Travel Using Laser-Pushed Lightsails" by R. L. Forward
1984. *Journal of Spacecraft*, vol. 21, no. 2, pp. 187–95.

15. "Magnetic Sails and Interstellar Travel" by D. G. Andrews and R. Zubrin 1988. Paper presented at a meeting of the International Aeronautics Federation, IAF-88-553.

16. "Starship Sails Propelled by Cost-Optimized Directed Energy" by J. Benford 2011, posted on the arXiv preprint server at http://arxiv.org/abs/1112.3016.

17. *Starship Century: Toward the Grandest Horizon*, ed. by G. Benford and J. Benford 2013. Lucky Bat Books. This book represents the proceedings of a conference by the same title in 2013, featuring scientists such as Sir Martin Rees, Freeman Dyson, Stephen Hawking, and Paul Davies, and science fiction authors such as Neal Stephenson, David Brin, and Nancy Kress.

18. *Frontiers of Propulsion Science* by M. Millis and E. Davis 2009. Reston, VA: American Institute of Aeronautics and Astronautics.

19. The 100 Year Starship (100YSS) project was started in 2012 by NASA and the US Defense Advanced Research Projects Agency (DARPA). The website is http://100yss.org/.

20. "Light Sails as a Means of Propulsion" by T. Dunn, unpublished calculations, online at http://orbitsimulator.com/astrobiology/Light%20Sails%20as%20a%20means%20of%20propulsion.htm.

21. "SpiderFab: Process for On-Orbit Construction of Kilometer-Scale Apertures" by R. Hoyt, J. Cushing, and J. Slostad 2013, final technical report to NASA on project by Tethers Unlimited, NNX12AR13G, online at http://www.nasa.gov/sites/default/files/files/Hoyt_2012_PhI_SpiderFab.pdf.

22. "Life-Cycle Economic Analysis of Distributed Manufacturing with Open-Source 3D Printers" by B. T. Wittbrodt et al. 2013. *Mechatronics*, vol. 23, pp. 713–26. Also "A Low-Cost Open-Source 3-D Metal Printing" by G. C. Anzalone et al. 2013. *IEEE Access*, vol. 1, pp. 803–10.

23. "A Self-Reproducing Interstellar Probe" by R. A. Freitas 1980. *Journal of the British Interplanetary Society*, vol. 33, pp. 251–64.

24. The original work is *Theory of Self-Reproducing Automata* by J. von Neumann, and completed by A. W. Burks 1966. New York: Academic Press. See also "An Implementation of von Neumann's Self-Reproducing Machines" by U. Pesavento 1995. *Artificial Life*, vol. 2, no. 4, pp. 337–54.

25. NASA funded a program called Breakthrough Propulsion Physics for eight years. Led by Marc Millis, it resulted in several workshops and a dozen technical publications. The project website says that no breakthroughs appear imminent. It also has this cautionary note: "On a topic this visionary and whose implications are profound, there is a risk of encountering premature conclusions in the literature, driven by overzealous enthusiasts as well as pedantic pessimists. The most productive path is to seek out and build upon publications that focus on the critical make-break issues and lingering unknowns, both from the innovators' perspective and their skeptical challengers."

26. "Possibility of Faster-than-Light Particles" by G. Feinberg 1967. *Physical Review*, vol. 159, no. 5, pp. 1089–1105.

27. "The Warp Drive: Hyper-Fast Travel within General Relativity" by M. Alcubierre 1994. *Classical and Quantum Gravity*, vol. 11, no. 5, pp. L73–L77.

28. Synopsis at the official *Star Trek* website: http://www.startrek.com/database_article/realm-of-fear.

29. "Teleporting an Unknown Quantum State via Dual Classical and Einstein-Podolsky-Rosen Channels" by C. H. Bennett et al. 1993. *Physical Review Letters*, vol. 70, pp. 1895–99.

30. Alice and Bob are two commonly used placeholder names, particularly in the fields of cryptography and subsequently in physics. The practice started because it's more personal and appealing than talking about A and B. The first use was by Ron Rivest in *Communications of the Association for Computing Machinery* in a 1978 article presenting the first public-key cryptographic system. Rivest says the choice of names is not a nod to the 1969 movie *Bob & Carol & Ted & Alice*. Communication of tangled quantum states follows this tradition: "Alice wants to send a message to Bob . . . ." If a third or fourth participant is needed, they're called Chuck or Dan. Eve is used for an eavesdropper in cryptography, or the external environment in a quantum communication situation. And that is probably more than you need or want to know.

31. "Quantum Teleportation over 143 Kilometers Using Active Feed-Forward" by X. S. Ma et al. 2012. *Nature*, vol. 489, pp. 269–73. For a report on worldwide teleportation by a University of Tokyo group, see http://akihabaranews.com/2013/09/11/article-en/world-first-success-complete-quantum-teleportation-750245129.

32. "Unconditional Quantum Teleportation Between Distant Solid-State Quantum Bits" by W. Pfaff et al. 2014. *Science*, DOI:10.1126/science.1253512.

13: COSMIC COMPANIONSHIP

1. In any product of numerical terms, the product is as uncertain as its most uncertain component. Having measurements of the incidence of habitable planets in the Milky Way doesn't mitigate our almost complete ignorance of the terms that related to alien physiology or sociology. Evolution on Earth led to intelligence and the development of technology by one species, but natural selection doesn't predict this as a necessary outcome; to argue that it does is to fall prey to anthropic bias. The counterexamples are the many species that did not become noticeably more complex, or evolve large brains, after hundreds of millions of years of evolution.

2. An anthropocentric way to estimate L is to use the average of human civilizations on Earth. Doing this historically gives an average of 300 to 400 years. See "Why ET Hasn't Called" by M. Shermer 2002, in *Scientific American*, online at

http://www.michaelshermer.com/2002/08/why-et-hasnt-called/. It's also possible that there may be many civilizations that are unstable or evanescent, with low L, but some that are essentially immortal, with very large L. David Grinspoon has discussed the implications of this for the Drake equation; see *Lonely Planets: The Natural Philosophy of Alien Life* by D. Grinspoon 2004. New York: HarperCollins.

3. *Contact* by C. Sagan 1985. New York: Simon & Schuster. Sagan and his wife, Ann Druyan, wrote the outline for the 1997 film, which was directed by Robert Zemeckis.

4. *The Dispossessed: An Ambiguous Utopia* by U. K. Le Guin 1974. New York: Harper and Row. This novel was something of a breakthrough for Le Guin, earning her literary recognition as well as the Nebula, Hugo, and Locus Awards for science fiction.

5. "Nikola Tesla and the Electrical Signals of Planetary Origin" by K. L. Corum and J. F. Corum 1996. *Online Computer Library Center*, Document no. 38193760, pp. 1, 6, 14.

6. The Proxmire incident was described in "Searching for Good Science: The Cancellation of NASA's SETI Program" by S. J. Garber 1999. *Journal of the British Interplanetary Society*, vol. 52, pp. 3–12. Bryan's attack is described in "Ear to the Universe Is Plugged by Budget Cutters" by J. N. Wilford, in the *New York Times* on October 7, 1993, online at http://www.nytimes.com/1993/10/07/us/ear-to-the-universe-is-plugged-by-budget-cutters.html.

7. "That Time Jules Verne Caused a UFO Scare" by R. Miller, online at http://io9.com/that-time-jules-verne-caused-a-ufo-scare-453662253.

8. "Where Is Everybody? An Account of Fermi's Question" by E. Jones 1985. *Los Alamos Technical Report* LA-10311-MS, scanned and reproduced online at http://www.fas.org/sgp/othergov/doe/lanl/la-10311-ms.pdf.

9. More than fifty (mostly) plausible explanations for the "Great Silence" and the absence of contact are laid out in *If the Universe Is Teeming with Aliens . . . Where Is Everybody?* by S. Webb 2002. New York: Copernicus Books.

10. *Rare Earth: Why Complex Life Is Uncommon in the Universe* by P. D. Ward and D. Brownlee 2000. Dordrecht: Springer-Verlag.

11. There's a robust argument to be made for procrastination. In any field where the technology advances exponentially, the sum of all previous projects will be eclipsed by the next project. In astronomy, this was the case through the 1980s and 1990s as CCD detectors advanced in size and sensitivity so rapidly that each new survey greatly surpassed the one that preceded it. The same argument could be made currently for mapping genomes. The argument is facetious, and of course science progresses because scientists continue to try to advance knowledge without waiting for the better capability that's imminent.

12. Built in the early 1950s, the Arecibo dish is a formidable radio telescope. It's so large that it's not steerable; it just stares at a strip of sky that passes overhead. The dish is made of aluminum panels equal in area to a dozen football fields. The feed

that detects the radio waves is suspended above the dish by three towers the size of the Washington Monument. Frank Drake likes to say that the dish could hold 100 million boxes of breakfast cereal or all the beer drunk on Earth in a single day.

13. "The Great Filter—Are We Past It?" by R. Hanson 1998, an unpublished paper archived online at http://hanson.gmu.edu/greatfilter.html.

14. "Where Are They? Why I Hope the Search for Extraterrestrial Intelligence Finds Nothing" by N. Bostrom 1998. *MIT Technology Review*, May/June, pp. 72–77.

14: A UNIVERSE MADE FOR US

1. *Year Million: Science at the Far Edge of Knowledge*, ed. by D. Broderick 2006. Giza, Egypt: Atlas and Company.

2. The evanescence of our civilization and cultural artifacts is sobering, given the technological prowess we exhibit. One book that conveyed this vividly was *The World Without Us* by A. Weisman 2007. New York: Picador. Weisman plays out a future where we cease to exist overnight and the infrastructure of human civilization decays and disappears with surprising rapidity. The Long Now Foundation swims against the dominant cultural trend by espousing "slower and better" over "faster and cheaper" and supporting projects with a millennial time frame. Most notably, its *Clock of the Long Now* is a mechanical timekeeping device designed to operate for 10,000 years without human intervention.

3. *Wired* magazine, April 2006, online at http://www.wired.com/wired/archive/14.07/posts.html?pg=4.

4. The experiments on mice are being conducted by Mark Roth at the Fred Hutchinson Cancer Research Center in Seattle. See http://labs.fhcrc.org/roth/. Dog experiments have been done at the Safar Center for Resuscitation Research in Pittsburgh. See http://www.nytimes.com/2005/12/11/magazine/11ideas_section4-21.html?_r=0.

5. *Cloning After Dolly: Who's Still Afraid?* by G. E. Pence 2004. Lanham, MD: Rowman and Littlefield.

6. "Embryo Space Colonization to Overcome the Interstellar Time Distance Bottleneck" by A. Crowl, J. Hunt, and A. M. Hein 2012. *Journal of the British Interplanetary Society*, vol. 65, pp. 283–85.

7. "Transmission of Information by Extraterrestrial Civilizations" by N. Kardashev 1964. *Soviet Astronomy*, vol. 8, p. 217. For his more recent work, see "On the Inevitability and Possible Structures of Supercivilizations" by N. Kardashev 1984, in *The Search for Extraterrestrial Life: Recent Developments*, ed. by M. G. Papagiannis. Dordrecht: Reidel, pp. 497–504.

8. "The Physics of Interstellar Travel: To One Day Reach the Stars" by M. Kaku 2010, online at http://mkaku.org/home/articles/the-physics-of-interstellar-travel/.

9. "Search for Artificial Stellar Sources of Infrared Radiation" by F. J. Dyson 1960. *Science*, vol. 131, pp. 1667–68.

10. "Fermilab Dyson Sphere Searches" using data from NASA's IRAS satellite, with results quoted online at http://home.fnal.gov/~carrigan/infrared_astronomy/Fermi lab_search.htm.

11. *Universe or Multiverse?* ed. by B. J. Carr 2007. Cambridge: Cambridge University Press. See also "Multiverse Cosmological Models" by P. C. W. Davies 2004. *Modern Physics Letters A*, vol. 19, pp. 727–44.

12. The first fine-tuning argument was the fact that the age of a biological universe cannot be too short or too long, "Dirac's Cosmology and Mach's Principle" by R. H. Dicke 1961. *Nature*, vol. 192, pp. 440–41. Since then, the idea has been explored by a number of physicists, for example: *Coincidences: Dark Matter, Mankind, and Anthropic Cosmology* by J. Gribbin and M. Rees 1989. New York: Bantam. Also: *The Goldilocks Enigma: Why Is the Universe Just Right for Life?* by P. Davies 2007. New York: Houghton Mifflin Harcourt. For a philosophical perspective, see *A Fine-Tuned Universe: The Quest for God in Science and Theology* by A. McGrath 2009. Louisville: Westminster John Knox Press.

13. "Naturally Speaking: The Naturalness Criterion and Physics at the LHC" by G. F. Guidice 2008, in *Perspectives on LHC Physics*, ed. by G. Kane and A. Pierce. Singapore: World Scientific. See also Prof. Matt Strassler's excellent primer online at http://profmattstrassler.com/articles-and-posts/particle-physics-basics/the-hierarchy -problem/naturalness/.

14. "Eternal Inflation and Its Implications" by A. Guth 2007. *Journal of Physics A: Mathematical and Physical*, vol. 40, no. 25, p. 6811.

15. *Impossibility: Limits of Science and the Science of Limits* by J. Barrow 1998. Oxford: Oxford University Press.

16. "X-Tech and the Search for Infra Particle Intelligence" by H. de Garis 2014, from *Best of H+*, online at http://hplusmagazine.com/2014/02/20/x-tech-and-the-search -for-infra-particle-intelligence/.

17. *Intelligent Machinery, A Heretical Theory* by A. Turing 1951, reprinted in *Philosophia Mathematica* 1996, vol. 4, no. 3, pp. 256–60. The von Neumann quote comes from Stanislaw Ulam's "Tribute to John von Neumann" in the May 1958 *Bulletin of the American Mathematical Society*, p. 5.

18. "Are You Living in a Computer Simulation?" by N. Bostrom 2003. *Philosophical Quarterly*, vol. 53, no. 211, pp. 243–55. The views of Kurzweil and Moravec are represented in their popular books, in particular *The Singularity Is Near: When Humans Transcend Biology* by R. Kurzweil 2006. New York: Penguin; and *Robot: Mere Machine to Transcendent Mind* by H. Moravec 2000. Oxford: Oxford University Press.

Credits

Figure 1 Creative Commons and Wikpedia/Ataileopard. **Figure 2** Courtesy Elsevier and Chuansheng Chen/University of California Irvine. **Figure 3** The scholar and academic skeptic Carneades, from the medieval book *Nuremberg Chronicle*. **Figure 4** NASA History Division. **Figure 5** *A Treatise of the System of the World* by Isaac Newton, published in 1728. **Figure 6** NASA Great Images. **Figure 7** Wikimedia Commons and Fastfission. **Figure 8** Wikimedia Commons and Lokilech. **Figure 9** Wikimedia Commons and Russian Federation. **Figure 10** Mark Wade/Astronautix .com. **Figure 11** U.S. Government/USAF. **Figure 12** Roel van der Hoorn/NASA. **Figure 13** NASA. **Figure 14** Wikipedia Commons and David Kring/USRA. **Figure 15** Wikimedia Commons and NOAA/Mysid. **Figure 16** Chris Impey. **Figure 17** Chris Impey. **Figure 18** Wikimedia Commons and Kelvin Case. **Figure 19** "Countdown Continues on Commercial Flight," *Albuquerque Journal*. **Figure 20** NASA/Regan Geeseman. **Figure 21** SpaceX. **Figure 22** NASA. **Figure 23** U.S. Government/FAA. **Figure 24** Wikimedia Commons and Nasa.apollo. **Figure 25** NASA/Kennedy Space Center. **Figure 26** Andrew Ketsdever. **Figure 27** NASA/JPL. **Figure 28** NASA. **Figure 29** Wikimedia Commons and Aldaron. **Figure 30** Matthew R. Francis. **Figure 31** Planetary Habitability Laboratory/University of Puerto Rico. **Figure 32** Postage stamp, Chinese State. **Figure 33** Wikimedia Commons and Dave Rajczewski. **Figure 34** Data source reports of Satellite Industry Association. **Figure 35** Patrick Collins. **Figure 36** NASA/Dennis M. Davidson. **Figure 37** NASA. **Figure 38** NASA/JPL/ University of Arizona. **Figure 39** NASA/JPL/Caltech. **Figure 40** NASA/John Frassanito and Associates. **Figure 41** NASA. **Figure 42** Christopher Barnatt/Explaining the Future.com. **Figure 43** NASA/MSFC/D. Higginbotham. **Figure 44** From *Xenology: An Introduction to the Scientific Study of Extraterrestrial Life, Intelligence, and*

Civilization by Robert A. Freitas, Jr., 1979, Xenology Research Institute, Sacramento, California. **Figure 45** Biosphere 2, College of Science, University of Arizona. **Figure 46** NASA. **Figure 47** NASA/JSC. **Figure 48** Javiera Guedes. **Figure 49** U.S. Government/LLNL. **Figure 50** NASA. **Figure 51** NASA. **Figure 52** Wikimedia Commons and Picoquant. **Figure 53** H. Schweiker/WIYN and NOAO/AURA/NSF. **Figure 54** NASA. **Figure 55** Wikimedia Commons and Fastfission. **Figure 56** Chris Impey. **Figure 57** Wikimedia Commons and Bibi Saint-Pol. **Figure 58** Andrei Linde. **Figure 59** Wikimedia Commons and Was a bee.

Index

Page numbers starting with 265 refer to endnotes.
Page numbers in *italics* refer to illustrations.